W. W. Walker

By northern lakes

Reminiscences of life in Ontario mission fields

W. W. Walker

By northern lakes
Reminiscences of life in Ontario mission fields

ISBN/EAN: 9783742876362

Manufactured in Europe, USA, Canada, Australia, Japa

Cover: Foto ©berggeist007 / pixelio.de

Manufactured and distributed by brebook publishing software (www.brebook.com)

W. W. Walker

By northern lakes

BY NORTHERN LAKES

REMINISCENCES OF LIFE IN ONTARIO
MISSION FIELDS

BY

REV. Wm Wm WALKER

Author of "An Itinerant in the British Isles."

ILLUSTRATED

TORONTO:
WILLIAM BRIGGS,
Wesley Buildings.

Montreal: C. W. COATES.   Halifax: S. F. HUESTIS
1896.

INTRODUCTION.

We have endeavoured as faithfully as possible to portray the following scenes and incidents, with a careful regard for the fundamental principles of truth, only recording those events which are distinctly and accurately remembered, spicing them up with stories of adventure by lake and forest. After all this painstaking labour, however, we are well aware of the fact that there will be imperfections and errors in connection with the work. But we crave the indulgence of the reader, and trust that he will throw the veil of charity over all, even though it does not reach his standard of literary excellence.

It is the earnest hope of the author that within the pages of this little volume there will be found something which will create a greater sympathy for those who have been called to battle with the difficulties of pioneer life in the northern districts of our country; also coupled with this, truths that we trust will lead some at least of our fellow-creatures to the exercise of a more steadfast faith in Almighty God, as they trace portraiture of exalted character, and learn something of the power of the Spirit in saving men.

The aim of the writer will be still further conserved if, through the reading of facts herein recorded relative to the almost illimitable resources, vast extent and gorgeous scenery of this mighty Dominion, Canadians are stimulated to a nobler patriotism, and to a profounder love of this their country, which is the fairest gem in Britain's world-wide constellation.

CONTENTS.

PART I.

CHAPTER I.—Arrival at Huntsville Mission—Rev. R. N. Hill—An unexpected conversion—Meet a bear on the road 9

CHAPTER II.—A prize-fighter and a singing dog—A perilous canoe trip—Wolves—A gracious revival—The prize-fighter converted 25

CHAPTER III.—Hunters' stories—A still hunt for deer—The missionary narrowly escapes drowning—Mr. S. and his Jersey cow—Hardy sons of Canada—Revival at the R—— appointment—Eloquent local preachers 36

CHAPTER IV.—A field meeting—Rev. Richard Clarke—A terrific storm—Wealth of timber and quartz—Canada's splendid heritage—Beauty of the northern lakes 50

CHAPTER V.—Preaching to Presbyterians—An Anglican associate—Sectarian bigotry deplored—A local geologist—Culture amid rude surroundings . . 60

CHAPTER VI.—A memorable hunting exploit—Salmon fishing—A chagrined hunter—Nearly swamped—We capture a fine buck—A toilsome march—Nature as a teacher 69

CHAPTER VII.—Revival at the F—— appointment—Service in a lumber camp—Disorderly conduct—An unexpected ally—Charm of the northern scenery—A hunter's encounter with a moose 80

PART II.

CHAPTER I.—Removal to the Parry Sound District—Intricate navigation—Magnetawan—Ahmic Harbour—Arrive at Dunchurch, our new home—Building and gardening 95

CHAPTER II.—Opening up a new appointment—Interview with a sceptic—Perils of bad roads—A distinguished American visitor—His sermon and war memories 103

CHAPTER III.—Fishing in the Magnetawan—Wonderful reclamation—Special services at D—— appointment—A fruitless chase—A hunter's lively encounter with a bear 113

CHAPTER IV.—A missionary tour of the lumber camps—Huge timber resources—A "character" in the settlement—A trying meal—The transforming power of grace 126

CHAPTER V.—A revival at F—— appointment—Devices of the enemy—Encounter with a drunken man—A murderous design frustrated 138

CHAPTER VI.—Two families portrayed—An example of true godliness—A contrast—A delightful sail—Camp-meeting at Parry Sound—Our Indian friends 144

CHAPTER VII.—Visit from the late Rev. Dr. Shaw—An entertaining guest—Dr. Shaw joins us in a hunt—Proves himself a good marksman—"A Christian gentleman" 152

CHAPTER VIII.—Area and material progress of the Dominion—Robust Canadianism—Trade statistics—Our new Premier—Canada's manifest destiny . . 163

BY NORTHERN LAKES.

CHAPTER I.

In the summer of eighteen hundred and eighty-eight the writer was appointed to take work under the chairman on the H—— Mission, in Muskoka, and in a few days proceeded by train to the town of Huntsville, where we were directed to the comfortable parsonage, then occupied by Rev. Richard Clarke, who presided over the District. On our knocking at the door an old and venerable-looking minister appeared, who looked us over from head to foot, and then cordially invited us to be seated inside. After an interesting conversation with this veteran of the Cross we partook of an excellent supper, and desiring to spend the night on our new charge, commenced to make preparations for walking three or four miles to the home of our nearest official; but Mr. Clarke would not hear of this, and compelled us to remain until he harnessed his horse and hitched to the buckboard, and then drove us to the home of Mr. T., where we desired to spend the night. After we had settled down, our host said, "How in the world did you come to be driven here to-night? This is the first time Mr. Clarke ever did that with a student." Then we remembered that the first question the Chairman had asked us was, "What country are you from?" and we replied that we were a Canadian by birth. "Then," said he, "what countryman is your father?" and we at once said a North of Ireland man. "Oh! well," he said, "you are the next best thing anyhow." And from that day forward we never lacked a friend while this estimable man lived.

After a night's rest at Mr. T.'s we were driven some miles farther to the home of another official, and having never but once seen a yoke of oxen previous to this time, were very much discouraged to find that every vehicle we met was drawn by cattle, and rocks on rocks skirted the road and adorned the landscape everywhere we turned. We had commenced a spirited conversation at the outset, but soon relapsed into utter silence through sheer disappointment at the appearance of things generally.

On arriving at the home of the other official, who, by the way, was the Rev. R. N. Hill, supernumerary, we were so kindly and hospitably received and entertained that we soon forgot, for the present at least, the poor and uninviting appearance of the country. Mr. Hill was a fine conversationalist, and informed us of the singular circumstance which led to his coming to Muskoka eighteen years before. Having a large family of boys, and finding it impossible to educate all of them, he resolved to go to some northern district where were free grants of land, and take up several hundred acres, and thus be enabled, when his sons desired it, to furnish them with farms, upon which they could make a comfortable livelihood; but one night before starting in quest of some suitable place for this purpose he dreamed that on the shores of a certain lake, which he had never seen, was the land he desired, timbered with trees as yet untouched by woodman's axe.

Mr. Hill at once prepared for starting, and went to the residence of a surveyor who lived within a few miles of the lake, and having explored the country around the point specified, he drew a plan of the waters and surrounding shores. But our host, who had never seen the place, except in his dream, corrected the surveyor, who had been on the ground more than once. Finally they both agreed to investigate the matter, and, procuring a canoe, after a long paddle and after portaging for some distance, arrived at their destination and found that the bays and points were located exactly as Mr. Hill described them. He at once regarded the matter as providential, and took up seven or eight hundred acres, which have proved to be the freest of rock and of the most productive soil in the entire district.

The above gentleman founded the Hillside Mission, which was named after him, and through those weary years, while the country was still a wilderness and but sparsely settled, did a mighty work for the Church of God by laying broad and deep her foundations in nearly every part of the district.

It must be remembered that when Mr. Hill settled on Peninsula Lake, Huntsville did not then exist; but on the site of what is now a beautiful and prosperous town there stood in lonely isolation a little log cabin with roof of wooden scoops. During our sojourn at the home of this retired and yet active minister, which continued for three weeks, before we settled on a boarding-place for the year, he entertained us with some very racy accounts of his experiences in preaching among the early settlers. On one occasion, just when he had announced his text and commenced his discourse, two dogs, belonging to members of the congregation, that had been left outside, had scrambled under the little school-house and commenced fighting, and so furious did they become that in their plunges they threatened to displace the floor, which was not very securely laid, and conclude the affair in the presence of the congregation, every member of which was terribly excited. One brother got up and stamped on that portion of the floor under which they fought, another yelled at them with all his might, and the scene became so ridiculous that Mr. Hill at last said, "Brethren, the good Book says without are dogs, let us leave them where the Bible places them and proceed with our service."

This able and useful servant of God, who, during the past year, was selected by the patrons of husbandry as their candidate for the Dominion Parliament, and who had done so much for the Kingdom of God in his community, passed suddenly away from his earthly labours to eternal reward, but his works do follow him. The circumstances in connection with Mr. Hill's death were very distressing. His second son was taken ill with inflammation of the lungs, and after days and nights of fearful suffering, during which the father never left the room for a single hour, was called from earth away. It was a crushing blow, and to complicate matters the steamer which was to carry Rev. Mr. W—— from the neighbouring town to officiate at the funeral, through some mischance never appeared, and that excellent and faithful servant of God, who was always ready to render help and minister consolation to sorrowing fellow-creatures, was, through no fault of his own, unable to be present. The father, however, with Christian fortitude, endeavoured himself to conduct the burial service, but broke down in the midst of it. Next day he visited the cemetery and selected his own grave, and on returning home was at once taken sick. During his brief illness he comforted those who were around him, and to his sorrowing wife said, "Do not weep, Caroline, it is scarcely worth while, for you will soon join me beyond the river." The last words he uttered upon earth were, "Blessed are the dead which die in the Lord from henceforth: Yea, saith the Spirit, that they may rest from their labours; and their works do follow them." After which, with a smile upon his countenance, he calmly fell asleep.

The Rev. Robert Norton Hill was a man of princely presence, standing six feet high, and weighing between one hundred and seventy-five and two hundred pounds, straight as an arrow, with over-hanging brows and of dignified manner. Our readers will readily understand, after coupling the above with undaunted courage, that as a magistrate he was a terror to evil-doers. But we may also add that he was a delight to those who did well, possessing a mellow spirit and sympathetic nature, which was evidenced by his aversion to the use of the gun. He informed us that on one occasion many years before, while out shooting he fired at a little bird, breaking a leg and crippling a wing. On proceeding to where it lay and witnessing its agonies he at once returned home, put up his "shooting iron," and never again used it throughout the remaining years of his life.

Mr. Hill was of good ancestry, also having inherited a considerable fortune from his grandfather. One member of his father's family is a distinguished minister of the Methodist Episcopal Church in the United States, another has attained to eminence in the medical profession—giving some conception of the intellectuality of

REV. ROBERT NORTON HILL.

the family—and the member of whom we now write was no exception to the prevailing rule; an able speaker, a close reasoner, with a mind well stored, not only with an intelligent knowledge of theology, but also with more or less of geology and classics, had he lived to enter Parliament, doubtless he would have made his presence felt in our halls of legislation, for his knowledge of politics and of the needs of the country was not insignificant. A beautiful shaft of polished granite marks his grave in Hillside Cemetery, erected to his memory by loyal and devoted sons, who keenly felt the loss of a wise counsellor and a loving father.

After the three weeks of our sojourn at the comfortable home of Rev. Mr. Hill, we secured board and lodgings for the year at Mr. R.'s. The house, though constructed of logs and small, was very cosy, and the inmates spared no pains to make the new missionary as comfortable as possible; indeed, we feel bound to use still stronger language and say, that never was mortal man used better than was the writer by this kind-hearted man (who was our Recording

Steward) and his estimable wife for the period of two years, which completed our labours in that part of the Master's vineyard.

After getting settled at our new home we commenced pastoral visitation among the people, and everywhere found them warm-hearted and hospitable, and also as a rule intelligent. Of course there were some exceptions, and ignorant people were sometimes encountered; but it will be remembered that this is the case in some of our most advanced districts. We admit the fact that schools are not convenient in certain localities, and because of this the education of the children is sometimes a little neglected; but with regard to native intelligence they are not below par. Also relative to the adult settlers, we met many who had received a liberal education, and who were right up to date in every respect; indeed, the fact of people going to a northern district like Muskoka when it was a wilderness to carve out homes for themselves and families, proved that they were enterprising and ambitious, and by no means the worst class of society.

On preaching around our new mission we found that there was but one church, and that constructed of hewed logs, and so small that it would not seat more than sixty or seventy persons; the other preaching places were little log school-houses. But at all of these we noticed that the people who gathered for worship were very attentive and well-behaved, and it seemed as though the outlook was most hopeful for sowing the seed of Gospel truth where it would bear fruit generously unto the honour and glory of the eternal God.

On one occasion we had preached in the only church on the charge, and the text had been, "Why stand ye here all the day idle," and a greater failure we thought we had never experienced. So great was our humiliation that we very much regretted there was not a door in the back end of the church through which we could get out without meeting any of the congregation; but the only means of egress was by the front, compelling us to face the inevitable and meet the people, whom we thought would be as much disgusted as the missionary with the sermon. But to our surprise an old lady was waiting at the stile, who, upon our approaching, took us by the hand, and with tremendous pressure and the tears raining down her cheeks, informed us that although she had been led to Christ thirty years before, and had fallen away into sin, and for many years revelled in it, yet she had during the service just held been reconverted, and never for a quarter of a century had she experienced such happiness. This is one of the evidences that it is impossible for a preacher to judge his own work or calculate the amount of good that is done by simply holding up a crucified Redeemer, though he may feel himself that the effort has been a failure.

While riding on horseback from the above appointment to the evening service in the little schoolhouse at H——, we were under the necessity of passing through five or six miles of woods, and in doing so noticed what we thought was a large Newfoundland dog slowly walking along the road. As we drew nearer we observed that the animal was minus the caudal appendage, proving at once that it was not a dog, but a large black bear. Knowing that, like all animals of its species, it was cowardly except when wounded or driven to bay, we at once put spurs to our horse and galloped towards it. Bruin hearing the pattering of hoofs at once commenced running, but we gained slowly, and as his bearship noticed this condition of affairs he soon sprang aside and disappeared in the woods. But we had got within about thirty yards of him and consequently had a good look at the first bear that we had ever seen in a wild state.

In our missionary expeditions here and there over a large tract of country we frequently saw deer upon the shores of the small lakes that abound in our northern districts. Sometimes we would see them standing in the middle of the road, where they would remain until we had got within pistol shot, and then the graceful creatures would bound into the woods. On one occasion, as we were passing through a wooded tract, there was a sharp report right ahead of us on the road, and as we reached the place indicated by the crack of the rifle, to our surprise a noble deer lay dead right across the track.

He had been standing in the roadway and was shot by a young man of one of our congregations at a range of about two hundred paces. Many of the Muskoka young men are so expert with the rifle that they can drop a deer at any distance up to five hundred yards.

CHAPTER II.

As the time advanced the writer saw evidences of the presence of the Holy Spirit in the Sunday services. During one of these a bright young man was convicted of sin, and accepted the Saviour. The text on the occasion was, " Whosoever will may come." We discovered that practising shooting with the boys was a pastime productive of much good. It called out their sympathy for the preacher, who they felt was one of themselves, and consequently they were very regular in their attendance at worship, and many of them, at a latter period, embraced Christ as their portion and became devoted followers of the Man of sorrows.

Some very amusing things transpired occasionally in our meetings. At one appointment called Antioch, situated on the extreme northern limit of the mission, we had recommenced preaching, it having been dropped the previous year. The place in which service was held was an old deserted house with broken windows patched up with shingles and rags. A dog followed some of the settlers into the building, and as they, on the commencement of the service, rose off the rough planks upon which they had been seated to engage in the singing of the first hymn, the canine, who was doubtless of a pious turn, also commenced singing or howling, and a man who was sitting on the first plank, and who was one of the greatest fighters and swearers in the Dominion of Canada, looked very much annoyed. However, we all thought the animal would soon quiet down, and we would not disturb him; but as the singing progressed our friend, the dog, also warmed up, and at last Mr. W., the prize fighter, got worked up beyond control, and springing from his seat caught the devoted dog by the back of the neck, and amid the most frightful howls ran across the room with him, and when he reached the door raised the hand that contained the now frightened animal and gave him a tremendous kick, sending him several yards, and at the same time advising him to proceed without delay to a much warmer climate than Muskoka. The congregation, which consisted of fifteen or twenty persons, broke down singing and burst out laughing, and we had to catch our flesh and pinch until we almost yelled with pain to keep back the nearly irrepressible volume of mirthfulness. In a moment or two, however, all again regained their equilibrium, and the service proceeded as usual, with our friend W. looking as grave as though nothing had happened.

After the memorable Antioch service just described we were invited by the brother who had ejected his dogship to remain to dinner, which we gladly did; and as we had to proceed several miles to an afternoon appointment he, in the kindness of his heart, suggested that his son paddle us across Sand lake, which would shorten the distance considerably. To this we at once consented, and got along famously until passing a certain point where the west wind had full sweep down the lake. Then we discovered it was very rough, and having only a small log canoe were in some danger, but lest they should think we were cowards we would not turn back, but paddled right on into the waves. The farther we went the higher they became, and perhaps the reader can form some conception of what it means to kneel in the bottom of a dugout, so tottlish that one would almost require to have his hair parted in the middle with precisely the same number of strands on one side as the other, lest it cause the canoe to capsize. Every advancing wave dashed over the bow and drenched us through and through, and had it not been for a tin dish which we fortunately possessed the people in N—— schoolhouse would never have heard the Gospel that afternoon, and woful tidings would have been borne to our friends in the south. But what with the remarkable coolness of the hardy son of the north who was paddling, and the persistent use of the tin dish in baling out the water, we succeeded in reaching the shore in safety, and without a moment's delay proceeded to the preaching place, where the congregation had been waiting about half an hour wondering what had happened. When we had concluded the service a little pool of water stood on the floor behind the desk where we had delivered our sermon, it having dripped off our soaked clothing.

Shortly after this experience on the lake we started Wednesday evening prayer-meetings at N—— appointment, and having no horse this year of which we write, walked seven miles from H—— every week to conduct it and returned to our lodgings the same night, making a walk of fourteen miles every Wednesday evening.

We had some peculiar experiences in returning late at night from the meetings mentioned above. Once the howling of a pack of wolves could be distinctly heard not much more than a mile away, and as we had miles of woods to traverse, and carried no other weapon than a light walking cane, you may readily conceive that we did not consider it the sweetest music that ever fell upon our ears.

On another occasion we almost stumbled over an enormous black bear in the darkness. We do not know who was most startled, but the bear did all the running, as we could tell by the report of breaking limbs and crackling branches, as he fled into the forest depths.

It was a singular fact, that after the first two or three meetings in the little log school-house there was a heavy rain storm every Wednesday evening for seven weeks. But we pushed through rain and everything, and always found someone waiting for the prayer service. After braving the storm for several nights we at last grew quite rebellious, and could not understand how it was, that when we were straining every energy to lead men to Christ, it seemed as if the very elements had been let loose upon us. At last we could stand it no longer, and on the seventh wet evening told the three or four young people who were present that we were completely discouraged, and could not understand why these things were permitted. But at the close of the service, to our infinite joy, a fine young man stood up for prayer, and as he had always been indifferent to his spiritual interests, his move in the direction of Christ and salvation caused considerable comment in the community, and we urged the people to come out on the next occasion that we might put forth a renewed and mighty effort for the conversion of those in sin. What with the appeal and the start made by the young man, and the fact that the next evening was fine, the schoolhouse was almost full, and after a rousing meeting two others stood up for prayer. Thus a revival had actually broken out in these services which we thought were going to be a total failure.

After the above manifestations of the presence of the Holy Spirit we at once announced for a service every night, and for two weeks laboured diligently for the salvation of men.

Every evening the little log school-house, almost surrounded by the dark woods, was crowded to the door with eager worshippers, and remarkable scenes transpired within its walls.

One night during the last week the old man who had led so rough a life, and who had been a terrible fighter, best known to us as the one who caused such a sensation at Antioch by the forcible ejection of the singing dog, leaped to his feet shouting, "Hallelujah, and glory to God! I've found it; I've found it; the pearl of greatest price." It was one of the most wonderful cases of conversion that we had ever known, and so great was the effect of it upon our meeting, that many others professed a change of heart the same night.

The news of Mr. W.'s acceptance of Christ spread like wildfire over the entire neighbourhood, and next night, though the writer was on time, he found it almost impossible to force his way through the mass of people that filled the aisle; they seemed to have come on this occasion to give themselves to the Master, and people who had been present afterwards told us that so great was the power of the Spirit, it was almost impossible to successfully resist.

Mr. W. had a neighbour who was a very bitter enemy, and who had received more than one chastisement at his hands, and who was present on the occasion of his conversion. Immediately on the close of the service, running up to the new convert and thrusting his clenched hand into his face, he called him a hypocrite and a deceiver, and a good many other things that are unmentionable. But the man who one short hour before would have instantly killed him, turned around meekly, and said, "Get thee behind me, Satan." Thus evidencing at the very outset that he had received a complete change of heart.

At the end of two weeks the services closed, after about twenty-five persons had professed to have found peace; their ages ranging from fifteen to sixty years. This number included nearly all the unconverted, or more correctly, those who had been unsaved in the community, so that almost the entire settlement was won for Christ.

Mr. W., in spite of the fact that his life had been so antagonistic to Christianity, became a zealous and devoted follower of his lamb-like Lord, and a faithful and loyal friend to the missionary.

In summing up the results of this peculiar revival, we are compelled to say, in all humility,

"That it is not by might or by power, but by my Spirit," saith the Lord. We need scarcely record the fact, that for a long time after this outpouring of the Spirit great interest was manifested in the religious services at the now noted school-house. Of course some of those who had been brought in lapsed, but most of them faithfully endeavoured to follow in their Saviour's footsteps, as long as we were associated with them.

One high-spirited young man was greatly incensed against the writer because some darling sin was exposed in the delivery of a sermon. We, however, knew nothing of the particular sin to which he was addicted, but the ground, through accident or the prompting of the Spirit, was completely covered, and in consequence the preacher was interviewed in the woods on his way to another appointment. The person in question said to us, "Did you intend that sermon for me?" In reply we said that if it fitted it was by all means intended for him, and furthermore, we were pleased that the Gospel had applied to his condition. He then said that he would never attend church again; but by the exercise of a spirit of charity toward him, he soon recommenced attending religious service, and at last became quite interested; he was unusually bright and intelligent, and we respected him very highly.

CHAPTER III.

In the intervals between the special efforts that were being put forth for the salvation of men, we were in the habit of indulging in occasional deer hunts. On these expeditions we were usually accompanied by some of the members of the church, sometimes by some of the officials, who were nearly all practical and skilful hunters. Many strange stories were related by these men of their early experiences, when the country was but sparsely settled; and as they were all reliable persons, these stories were intensely interesting. One of them on a certain occasion had been chopping and was returning in the evening in his canoe, as he had to cross a small lake to and from his work, when he noticed a deer in the water which was swimming towards the opposite shore. He instantly gave chase, gaining upon the animal at every stroke

THE WHALEY LUMBER COMPANY'S MILL, AND RESIDENCE OF THOS. WHALEY, ESQ., HUNTSVILLE.

of the paddle, and at last coming up to within striking distance, rose to his feet to administer the death-blow with his axe, the only weapon in his possession. But, as he struck with all his might at the animal's head, the birch-bark canoe shot away from under his feet, and he was precipitated into one hundred feet deep of water. Suffice it to say that the deer escaped and the hunter was nearly drowned.

On one occasion Mr. N. and Mr. H. of the N—— appointment, invited the writer to accompany them on a still hunt for deer; that is, stalking them without dogs. We readily consented, and as the snow had fallen a foot deep the previous night it was considered a very favourable time for our diversion. Accordingly next morning we started, and after a toilsome march through the deep new fallen snow, reached the shore of a small lake, and as it had frozen hard for two nights, our companions thought that it would be perfectly safe to cross on the ice, as it would materially lessen the distance to our hunting ground.

Though not particularly liking the idea of

crossing the lake in this way, we did not wish to be obstinate, and acceded to the desires of more experienced woodmen. As a precautionary measure, however, and not desiring to push ourself forward, or to appear bold, we allowed the others to go first, as they represented more matter avoirdupois than the missionary, one of them weighing over two hundred pounds. But this modesty that preferred others before self was nearly proving our ruin, for the two heavier men, on passing over the thin ice, cracked it, and after proceeding a short distance the poor preacher, loaded down with a heavy Winchester rifle, together with forty rounds of ball cartridge and a day's rations, broke through the weakened ice, and in much less time than it takes to record it, went right to the ears in a place where the water was thirty feet deep.

In this dilemma our companions could render no assistance, for if they attempted to approach the hole it would probably seal the fate of the entire party. The rifle, eventually, proved our salvation, for getting it above the sound ice we were able to pry ourself out of the water, and fortunately in doing so the edge of the good ice did not give way, and by rolling over a few times (we dared not stand up, lest we again go through) finally succeeded in getting to the shore in safety, where our clothes were soon frozen like planks upon us, it being an intensely cold day. As it was noon we thought it advisable to kindle a fire for the double purpose of warming our chilled bodies and thawing out our frozen food. After partaking of our sandwiches, which we had literally to toast in order to get the frost out, we struck into the woods instead of again attempting to cross the lake, none the less determined, for our accident, to make the hunt a success.

After an exhausting tramp of many miles through the tangled underbrush, and also through the new fallen snow, we reached a place where deer tracks were very numerous, and at last impressions made by the bodies of the animals were distinctly visible upon the snow. We now scattered in different directions, each one of us springing a cartridge into the breech of his rifle and moving at "the ready," expecting a shot every moment, but no deer appeared, and after moving a considerable distance in this way we signalled each other by discharging our firearms, and once again got together, starting instantly for home. On the way we suffered considerably from the difficulty of locomotion over high hills, their sides in some cases covered with rock, in others with dense underbrush. At last, however, the monotony of the march was broken by the sight of what every one of us declared was a large antlered deer. The object was about two hundred yards distant, and, after a short consultation held in a whisper, it was decided that to advance would attract his attention, and we would probably lose him, as a flying shot is always doubtful; so raising the elevations for the present range two shots were fired. But, to our infinite amazement, the object never moved, not even to topple over. Another shot was tried and still no motion. Unable to account for this we at once went forward to investigate, and to our chagrin found that the object at which we had fired three times was the root of a tree, which had its edge turned toward us, and looked more like a deer than anything we had ever seen, deceiving two experienced hunters, as our companions were. The only consolation, however, in connection with the matter was that all three shots had taken effect, proving that had it been a living animal its doom was sealed.

Old hunters, with all their experience, sometimes make calamitous blunders along this line. One of the best in the entire district, who had killed hundreds of deer with his rifle during the fifteen or sixteen years of his sojourn in the north, had purchased a Jersey cow at a cost of one hundred dollars, which animal was allowed to run in the woods with the other cattle. One day, however, Mr. S. was out hunting and noticed an object moving in the brush, which he at once mistook for a deer and fired instantly, at a range of one hundred and thirty yards. The animal gave one wild plunge forward and fell. Mr. S. at once ran forward, delighted that he would have a fresh supply of venison, but to his horror discovered that he had killed his thoroughbred Jersey cow, which was shot through the heart. Feeling very much ashamed, he resolved that he would not tell any of the neighbours; but in spite of every precaution the truth leaked out. Mr. S. said, however, that he always consoled himself with the thought that although the shot cost him so great a price no man could aim it better.

After the little adventure mentioned above we proceeded as rapidly as possible in the direction of home, which we reached in a very tired condition, about ten o'clock at night. Thus, after a march of about twenty miles through twelve inches or more of fresh snow, and after climbing many a hill with difficulty, with baptism by immersion gratuitously thrown in, and after utterly failing to secure any game, we were in a position, and also in a condition, to appreciate the glowing fire and hot supper which were so much enjoyed in Mr. N.'s log shanty.

During the hunt above recorded we had an excellent opportunity of seeing the hardihood, and noticing the courage and determination of those typical Canadians, who had been cradled and nurtured amid northern scenes. It was such men as these, of whom an American military officer said, "The Canadians who fought in the Federal ranks saved the Union on more than one bloody field. Everywhere their bayonets flashed in the front rank of armed courage, everywhere they inspired their American comrades by the example of an unconquerable valour."

If ever the soil of our beloved Canada is polluted by the tread of hostile invaders, bent on robbing us of our righteous laws, Christian liberty, or national honor, methinks those hardy sons of the north, with their splendid physique, brave as lions, and crack shots with the rifle, together with the frontiersmen of the sunny south, would hurl back, in consternation and ruin, the presumptuous despoilers..

Also, if the sons of Canada are true first to their God, second to their splendid God-given heritage, third to themselves, history will accord them an honourable and conspicuous place in recording the annals of the nations.

After the above experiences it was judged advisable to begin revival services at the W——

appointment, where was now a good congregation and unusual spirituality. We did not propose, however, holding them more than a week or two; but, after starting, we found so much interest manifested therein that it was practically impossible to stop, even at the end of two weeks; accordingly for a period of nearly one month the work went on, although the latter part was characterized by one of the most severe snowstorms ever witnessed; the roads were completely blocked and for two nights it was impossible for a single soul to reach the church. However, the third brought out almost our entire congregation once more. We did not witness the singular scenes in this place which were witnessed at the N—— school-house, the work was quietly done by the Holy Spirit, and as the result of the month's labours some twenty-two persons professed conversion.

It is scarcely necessary to say that as the result of these revivals the cause of the Master was greatly strengthened in this part of His vineyard.

The missionary takes no credit to himself for

the success of these special efforts, which was not only due to the ministry of the Spirit, but also to the intelligence and zeal of the officers of the church. This zeal, coupled with knowledge, heart consecration and Christian unity, was owned and blessed by Him from whom all blessings flow, thus starting an influence the power of which eternity alone can reveal.

For a considerable length of time after the last series of services the old routine was resumed, pastoral visitation, prayer-meetings and the ordinary Sunday services. As we were often called away from the latter to supply for the Chairman or to help some brother in difficulty, our place was admirably supplied by good and worthy local brethren. Mr. H. and Mr. S. were both zealous and eloquent preachers, and were most acceptable all over the mission. The latter was in many ways a remarkable man, being familiar with botany, natural history, science, philosophy, mathematics and many other subjects. He had in his library about one hundred volumes of literature, which in his humble little home was rather a surprise. Having two boys, the eldest of whom was about fourteen years of age, he spared no pains in instructing them. Attending school was out of the question, as the nearest was miles away; but those boys, who had scarcely ever seen the inside of an institution of learning, could talk about cellular tissue, protoplasm, and primordial utricle, together with the philosophical laws of association and suggestion, like a college professor.

The attainments of these youths teaches an important lesson, that it is possible, with the possession of a good mind, even though one has never attended a school, to become a critical scholar. Their knowledge of literature in nearly all departments was truly astounding.

Mr. H. had not read as extensively as Mr. S., but he had a natural gift of oratory, and in the delivery of his sermons would become quite eloquent. This inherent qualification atoned in large measure for his not having studied much, and made him popular as a preacher. We have listened to him when his heart was warmed and his soul enthused with the subject in hand, when his outbursts of eloquence were amazing. We often thought that he had in him the elements of true greatness.

During our ordinary pastoral work we had an excellent opportunity of studying human nature and character, and also of judging the intelligence of the people, which in some cases was remarkable. Without academic training, without social advantages, without books, they had, by close observation of men and things, attained to a degree of culture that would surprise a southerner.

Also, with regard to morals we might add that in this northern clime, what with the cold, bracing atmosphere, plain, wholesome food, and not too much of it, and incessant hard work, the people have little disposition to the commission of gross crime; and in all our labours in the north we rarely heard of a case of trangression of the law of social purity. This proves that apart from Christianity the best safeguard along this line is temperance and diligence in business, on the principle that "Satan finds some mischief still for idle hands to do."

CHAPTER IV.

AFTER considering the matter, and consulting some of the brethren concerning it, it was judged advisable soon to devote an entire Sabbath to the holding of a field meeting for the benefit of the whole mission. The idea was so novel that many people who seldom attended church at once fell in with it, thinking, doubtless, that it would be a species of diversion. All other appointments were accordingly withdrawn, and on the arrival of the longed for Sunday a large number of people gathered at the appointed rendezvous, a place that had been cleared of underbrush, on gently sloping ground, in a part of the woods which was convenient to the Government road.

The interest in the services was shown from the fact that many persons came a distance of six or seven miles, some in canoes by water, and

REV. RICHARD CLARKE.

some on foot, whilst one or two walked much farther. The attention and deportment, although so large a number were present, was all that could be desired.

The preacher of the day was Rev. Richard Clarke, Chairman of the Bracebridge District. The morning text was taken from the last verse of the seventy-third Psalm. "But it is good for me to draw near to God: I have put my trust in the Lord God, that I may declare all thy works." The text for the afternoon was taken from Galatians the sixth chapter and the seventh and eighth verses. "Be not deceived; God is not mocked: for whatsoever a man soweth, that shall he also reap. For he that soweth to his flesh shall of the flesh reap corruption; but he that soweth to the Spirit shall of the Spirit reap life everlasting." Both discourses were able and masterly to an exceptional degree. All who have known Mr. Clarke are aware of the fact that he was not only capable of manufacturing, but also of delivering, a first-class sermon. In his expositions of Gospel truth on this occasion there were many beautiful touches of eloquence that captivated his hearers. This was especially noticeable in the morning effort, when the venerable divine held his audience spell-bound for over an hour, his force and vigour and powers of oratory unimpaired by the lapse of years.

The last service was scarcely over when one of the most terrific thunderstorms that we had ever witnessed burst in all its fury upon us. Many of the people leaped from their seats panic-stricken. The benediction was pronounced amid the crash of falling trees, the livid flashes of lightning, seeming to rend the skies, being followed instantaneously by bursts of thunder that resembled the discharge of twelve-pounder field pieces. But strange to say, although the wind with cyclonic power tore enormous trees up by the roots and scattered branches in every direction, yet not a single individual was hurt, and scarcely half a dozen heard the Amen. Everybody was drenched through, our worthy chairman included. We at length, however, reached a neighboring house, and although all were dripping wet, yet universal good-humour prevailed. Nobody seemed to regret having come, and for nearly an hour a large number of people were huddled together, almost like herrings in a barrel, discussing the events of the day. As Mr. Clarke was announced to preach at H—— in the evening he now said that he must start for home. As there were still several hours before church time we earnestly solicited him to come to R——, but a short distance, and change his garments. He, however, absolutely refused and drove about eight miles in the chill damp of a May evening, preaching that night at H——, ever faithful to his trust. But he had taken cold from wearing his wet clothes during the long drive, and from that time forward his health gradually declined, finally culminating in his lamented death.

> "But though the Christian warrior's sun has set,
> His light shall linger round us yet,
> Bright radiant blest."

The appearance of the storm as it advanced through the woods was grand and awful in the extreme. It was most interesting as it drew near to watch the countenances of the people composing the audience. Some were blanched in terror, while others sat bolt upright, their faces indicating sublime indifference and a coolness worthy of war veterans. It seemed almost miraculous that the entire company escaped without some fatality, or some one being injured; yet such was the case, and altogether the day was one long to be remembered in the annals of the H—— Mission, as one of spiritual profit and of awful grandeur, as one from which influences are sure to start, of an elevating, of an ennobling, of an exalting character, which tend to draw men more closely together in the bonds of holy fellowship with one another, as well as into closer communion and fellowship with the triune God.

On the way home, those who went in their canoes found the small lakes dangerous to navigate because of the number of logs and branches of trees floating in their waters, one of the evidences of the fury of the wind. We always noticed that after a thunderstorm in the

North land the air was delightfully balmy. So evident was this, and so refreshing and invigorating the feeling from the delightful atmosphere, that there seemed to be a universal desire to quit the circumscribed limits of the houses for a draught of that, the fragrance of which no language can describe.

During the summer following our field meeting, we had in our journeys, in both Muskoka and Nipissing Districts, in connection with our regular work, an exceptional opportunity of studying the scenery, which is surpassingly beautiful.

Our readers will understand why we speak of journeying in another District, by the fact that a new appointment had been opened up some miles within its bounds, thus necessitating our making this journey every two weeks.

With regard to the hills of the Nipissing District, some of them were so high that they almost attained to the proportions of mountains, their sides clothed with living forests. Oh, what a wealth of timber is in those northern climes! It seems as though the supply is almost inexhaustible. And within those flinty walls of rock that tower up towards heaven there must be concealed mineral wealth sufficient, with proper development, to enrich a nation. Gold quartz has been found in several places, and doubtless other ores abound.

As Canadians we have scarcely yet comprehended the importance of our vast country, which is bounded on the north by the Arctic Ocean, on the east by the Atlantic, on the west by the mighty Pacific, on the south by a great chain of inland seas.

Yet gradually the truth dawns upon us in this closing decade of the nineteenth century, that what with the coal, timber and mineral deposits and limits of the bracing north, the productive wheat fields of the far west, and the fruits and cereals of the glorious south and east, the resources of our noble Dominion are almost unlimited.

Our majestic rivers, the St. Lawrence, Ottawa, Saskatchewan, Mackenzie and others, some of them navigable for a thousand miles, literally teem with a great variety of delicious fish. On their banks also are, on the one hand, beautiful farms and comfortable homes; on the other, during many months of summer, they are fringed with foliage, many hued, that sometimes in mirthful glee kisses the silvery waters, giving to the entire scene the aspect of a paradise.

Also our lakes cannot be forgotten. In surpassing beauty they bound and dot the land from Ontario's rippling waters to Superior's azure blue; and on to the little gems that adorn our northern clime, diademed by islets that in turn are crowned with lordly trees. On their shining bosoms ride fleets of magnificent palatial steamers, in their watery depths there is hidden food for millions, while on their shores are cities which remind us of Augustine's city of gold.

CHAPTER V.

In the autumn of this year some of our Presbyterian friends from D—— waited upon us, and as their student had returned to college, invited the writer to give them a sermon every two or three weeks, to help keep their congregation together until the term closed in the spring. Knowing those brethren well, and esteeming them highly as we did, we at once consented to preach on a week evening, every fourth night, at D——. This course we pursued through the fall and winter, and very much enjoyed showing this little kindness to our friends of a great sister denomination; and we record this fact that no man could have been better treated by his own people than was the writer by these kind friends of another denomination, some of whom were very intelligent—one, who was an elder in his church, taking as many as seven newspapers, and thus keeping himself fully abreast of the times. This gentleman had two sons, unusually intelligent young men, who were fond of reading and study, and who were greatly respected in the community. The eldest was a tall, fine-looking young fellow, who, most unfortunately to all human appearance, soon met a tragic death, being shot through the heart by the accidental discharge of his rifle while on a bear hunt.

We frequently met with Mr. L., the young Anglican student who had a charge at G——, and very much enjoyed his company. He had splendid social qualities, and in a short time we became quite intimate. Mr. L. had, though quite young, been a Volunteer for some time previous to his taking a charge, and consequently was an enthusiastic soldier. It was most interesting to listen to his amusing and comical stories of adventure during camp life. He was also in his regular work a good pastor, possessing a sympathetic nature, which, coupled with pleasing manners and good address, made him a welcome visitor everywhere. In spirit he was most liberal, preaching for us on one occasion at our Wednesday evening prayer-meeting and, we believe, greatly benefiting the people.

The above associations conspired together to broaden us so as almost to forget denominationalism, without generating disloyalty, and even the two or three Catholic families on our charge were regularly visited. They were always sociable and hospitable, thus making it a real pleasure to call on them.

It often has been suggested to us that if we associated to a greater extent with our Roman Catholic neighbors, the high walls of bigotry and prejudice that have in the past towered up toward heaven, would soon crumble into utter ruin, and we would discover that as surely as the fatherhood of God was an established and recognized fact, so surely also was the brotherhood of man; for the eternal Word teaches us that all men are brothers, and of the same origin. See Acts xvii., 26-28: "And hath made of one blood all nations of men for to dwell on all the face of the earth, and hath determined the times before appointed, and the bounds of their habitation; That they should seek the Lord, if haply they might feel after him, and find him, though he be not far from every one of us: For in him we live, and move, and have our being; as certain also of our own poets have said, For we are also his offspring."

Also see Romans xiv. 10: "But why dost thou judge thy brother? or why dost thou set at nought thy brother? for we shall all stand before the judgment seat of Christ."

Let us remember that the Catholics of this Dominion are our countrymen, that they have their homes and interests here just as we have, and that the same patriotism and love of country that wells up within us also fires their souls and fills their breasts.

We are not so much Presbyterians and Anglicans, Congregationalists and Baptists or Methodists; yea, more, we are not so much Protestants and Catholics, as we are Canadians. Let us therefore trample bigotry into the dust of earth forever, and cultivate a purely national sentiment founded upon the basis of eternal truth, recognizing the fact that we are not only fellow-citizens, but also that we are brothers.

The proper time may not have come for an organic union of the different sects of Christendom; but to those of us who are most loyal to our own particular Church the time has come for union in spirit, and no matter what our creed may be let us remember that we are "One army of the living God." At His command we must bow. Bigotry, which is always narrow, is not only an abomination to intelligent and thoughtful men, but is also highly displeasing to the eternal God who controls the destinies of men.

Among the many sects represented on our charge was one known as the Marshallites, and the most intelligent layman among them was Mr. W. His business was prospecting, and so long had he followed it—perhaps forty years or more — that he became a fairly clever geologist. He could classify the different rock formations with some degree of exactitude, and it was most interesting to listen to his explanations, from the various strata, of the possible age of the earth. He also very severely criticized Hugh Miller's "Testimony of the Rocks," proving him to have been absolutely wrong in some cases.

Mr. W. was well endowed intellectually, having a scientific mind of some power. Of course he very persistently adhered to his peculiar religious belief; and although brainy, thus proved himself to be not a little narrow in general scope. But, after all, he was undoubtedly a godly man and a devoted student of nature, behind which his keen mental perception enabled him to discern and love Nature's God.

We trust none will infer that we hold up all the inhabitants of our northern districts as an example of intelligence, for our readers may rest assured that Mr. W., of whom we have just written, was an exception, not the prevailing rule. There were few like him and Mr. S., whom we have mentioned in a previous chapter, to be found in any district of any province in our country, or, considering their opportunities, in any other land.

To prove that all were not so clever, an excellent opportunity soon occurred. One of our neighbours lost a child by death, and on the day of the funeral, for some reason which we cannot now recall, there was no service held in the house, but instead of this we called on two praying brethren at the grave to lead us in devotion, and they first thanked the Lord for such an opportunity of meeting together! The reader can imagine the effect of this in the presence of the grief-stricken parents. We do not now remember the denomination to which they belonged.

On another occasion a man was describing his visit to the neighbouring town to an intellectual friend, and of course tried to use choice language, knowing that such would be pleasing to a man of culture and refinement. But unfortunately the speaker was not a master of English, and did not know the difference between syntax and beer tax, consequently saying, "I went to town to expose of some prudence and was obtained by the rain."

An important lesson is to be learned from

this. If we are not thoroughly posted in the science or art of terminology, we had better be exceedingly careful as to the language we attempt to use. Better is it for an illiterate man to use plain terms than to make himself ridiculous by any attempt at the use of complex or abstruse phraseology. We shall proceed a step further; simple language is becoming to all, both learned and unlearned.

The way in which some have instructed themselves amid the wilds of the north is indeed astonishing to those of the south; making themselves men of broad culture, in many cases, amid surroundings which were anything but helpful. We are fully in sympathy with schools and the higher collegiate training, and any one who was not would be a dangerous enemy to society and the State. But experience in the Canadian mission field has taught us that it is possible to be thoroughly educated without ever having seen the inside of either school or college.

There are lessons in the towering rocks, in the forest-crowned hills, in the picturesque lakes, in whose rippling waters is mirrored the face of the Infinite One, that instruct and profit. As one whose character is recorded in ecclesiastical history has said: "The very beeches taught me oratory." There is a nameless something amid wild enchanting surroundings which leads the mind of man out in a longing after knowledge, the satisfying of which will eventually develop independent thought.

CHAPTER VI.

AFTER depicting men and things, together with varied scenes in the foregoing chapters, as faithfully as our limited ability would permit, we shall now proceed to record one of the most eventful experiences of a lifetime, when in company with three or four experienced hunters and canoe men we organized for the chase on Monday morning, October 14th, nearly eight years ago, and commenced the march. A yoke of oxen were brought along to carry canoes, provisions, etc., as far as we desired to go by land, and were then sent back in charge of a boy. It was rather early in the evening when we reached Tasse Lake, which in the past has had the reputation of being a great resort for deer. Immediately on reaching the shore of this beautiful body of water, hidden among the surrounding hills, and fringed by unbroken forest, we proceeded to construct a camp by placing a pole across the branches of two trees, and laying a few others obliquely against this, piling balsam boughs on top, also throwing a few inside to sleep on.

The deer season not opening until next day, the 15th, we fixed up our trolling lines and fished for salmon an hour or two in the lake; but as our canoe was simply a dugout—that is, a log scooped out with an axe—and had lain somewhere in the sun until seamed in the bottom—not sufficient to cause it to leak rapidly, but just enough to let in a couple of inches in an hour—we got thoroughly wet. We were forced to sit in the bottom while paddling, and having no changes of raiment, were compelled to attempt drying ourselves by standing beside the fire. Not being very successful in this, however, owing to the intense heat, we had to defer other efforts in this direction until the sun completed the toasting process next day, as we stood on our runways waiting for a shot.

The result of our fishing was not only to get wet, but also to catch a salmon weighing about a pound and a half. This, of course, was only a taste for several hungry men, yet we were pleased to get even that, and felt thankful for small mercies.

Failing to catch any more on this occasion, we retired to the seclusion of our camp and proceeded to cook supper. We might remark at this juncture that we did not observe all the forms and ceremonies of modern society in the cooking and serving of our meat, but went at it in true hunters' style, and enjoyed the whole proceeding immensely.

Being very hungry, after our long march through the woods, and after paddling around the lake in search of fish, we partook of each course with a peculiar relish. When our repast was finished we talked and joked around the camp-fire until about ten o'clock, when one after another dropped off, completely overcome with sleep, and soon we were all wrapped in dreamless slumber, covered with two old quilts, all that we possessed.

We were astir early next morning cooking our breakfast and partaking of it as rapidly as possible. After the meal we had prayer, a thing that rational creatures, say nothing about Christians, should never forget in any condition or under any circumstance. For on hunting expeditions, like military compaigning in war time, men are continually exposed to danger, and sometimes death lurks very near, by accidental shooting or the capsizing of a canoe. So we trust our readers will remember that it is always wise, no matter where we human beings are, to commend ourselves to the care and guardianship of the supreme and eternal God, who holds our destinies in the right hand of His power.

After our devotions were completed we at once proceeded to our stations on the runways leading into the lake. The dogs were now let loose and soon started a deer. He followed a trail leading to the narrows in the body of water above referred to, and coming to the edge coasted along quietly for some time, the dogs having got off the scent. The watcher opposite feeling sure that the animal would cross at this point—it being only fifty or sixty yards wide—refrained from firing, though the distance was so short that a marksman could easily have killed the deer with a good revolver. His deership, however, after leaping over a few logs on the bank, suddenly sprang into the woods and was out of sight before the hero who was concealed on the other shore had time to raise his rifle. His chagrin may easily be imagined, and he was continually reminded for a week afterward of his bungling on this occasion. At the end of that time, however, the teasing became so intolerable that he threatened to leave us and go home if we did not desist. Accordingly, being unwilling to lose him, as he was the best canoe man of the party, we dropped the matter.

We had another hunt in the afternoon of this day, but did not succeed in shooting anything, the game all going east on being started by the hounds. As the result of our failure in this place we struck camp next day and proceeded up Tasse Lake in quest of better ground.

We had a little log canoe, intended originally for but one person; but having only two altogether, and one of them seamed, as we mentioned previously, and four persons with their equipments, two of us with arms, ammunition and provisions, aggregating three hundred and forty or fifty pounds weight, we were compelled to embark in it. The gunwales were almost on a level with the surface of the lake; and as we proceeded on our way the tottlish concern shipped water several times, and having nothing with which to bale it out, it was soon half-full.

The occupants of our other canoe seeing the danger called to us to immediately put into an island that was near. We did so, and not any too soon, for when we had almost touched shore the dugout swamped under us, and we got badly wet. After emptying the contents of our frail bark upon the island's shore and transferring some of our accoutrements to the custody of the occupants of the other boat, we again started, soon reaching the head of those waters without further mishap. Then striking across a trackless wilderness, carrying all our equipments, after a tiresome march, we reached Marion Lake just before dark. Building a fire, we cooked our supper, and banishing all ceremony, devoured it like a lot of ravenous wolves.

Then lying down with mother earth for a couch, the heavens for a canopy, and the darkness for a winding sheet, and five miles from any human habitation, we slept and dreamed of comforts far, far away.

Next morning we rose shortly after daybreak, and partaking of our morning repast and repeating our matin prayers, we at once proceeded to our posts of duty, with the understanding that any member of the party who let a deer escape, that had come within easy range of his rifle, would be severely dealt with.

We lay in our places on the runways almost breathlessly awaiting coming events, and the morning being crisp and frosty—a typical chill, October one—our teeth occasionally rattled and our knees involuntarily smote together with cold. Soon, however, hearing the dogs give tongue (in hunter parlance) we shook off the chill and lay at the ready. We had not long to wait, for shortly after ten o'clock a large deer with splendid antlers leaped into the water with a splash that made us all jump.

Gracefully and beautifully did the antlered

dweller of the wild propel himself, swimming powerfully for the other shore. We watched him for a short time in bewildered admiration, then recovering ourselves and thinking it time to close the scene, opened fire on the noble animal, which, indeed, seemed cruel. The first shot struck a little behind him, the marksman miscalculating his speed. The second was better aimed, hitting him in the side of the neck and stunning him for a moment. He soon, however, recovered himself and swam more furiously than ever, having changed his course. The back of his head was now fully exposed to us, and a bullet from a repeater, aimed with deadly skill, broke his neck, his nose instantly sinking in the rippling waters. When we had towed him ashore we found that he would weigh in the neighborhood of two hundred and fifty pounds.

We had another hunt in the afternoon; the deer travelling in the direction of the East River, however, brought our plans to naught. In the intervals between the runs our Kennedys and Winchesters made sad havoc of the partridge, which made an excellent dish in the far-off wilderness. We had now come to the end of the pleasant part of our experience, for having eaten so much meat the last day or two, partly because our other provisions were running low, and partly because venison and wild-fowl were novel luxuries, we became surfeited with both, nearly all of us finally developing a loathing for anything in the form of flesh, so that when our supply of bread and potatoes was absolutely exhausted we began to starve. In no way daunted, however, next morning we were up at daybreak, and having nothing that we could eat except a few cakes, made from flour and water, we were compelled to strike the runways breakfastless, and did not return to camp until high noon, and then all the dinner which we had was a small piece of biscuit each.

Almost immediately after partaking of our frugal meal we made up our effects in packages of fifty pounds each, as nearly as we could guess, and started on the toilsome march towards civilization, each man with his bundle of half a hundred weight strapped on his back, and after considerable hardships and suffering from the pangs of hunger we reached the nearest settlement a little before dark, having traversed five weary miles, fording numerous creeks and plunging through almost impenetrable forest, and then urged by kind hearted settlers we endeavored to satisfy the craving of nature.

We were certainly a fierce-looking party when we reached the habitation of the nearest pioneer; what with torn clothes, unshaven faces, and armed as we were to the teeth, our friends would scarcely recognize us. But the experience was invaluable; and how can thinking men study God to better advantage than when revelling amidst the gorgeous scenery of mountain, lake and stream, which, with voiceless eloquence, proclaim the handiwork of Him who rules on high, and who, with a skill that at once stamped him the Master Workman, formed and fashioned it all.

So exhausted were we after the hardships of the expedition that it took us a considerable length of time to recuperate and settle down once more to regular work. We also learned that it is not sufficient to alone study the productions of the minds of great men who have lived, some of them in the far past, or to acquire merely a theoretical knowledge of things, or as mere book-worms to gratify longing desire for information in the seclusion of the study; but that in order to broaden our conceptions generally, we must take for our text-book the great volume of nature, and within its illuminated pages we will find, not false teaching or misconception and error plausibly set forth under the guise of orthodoxy, but the very epitome of truth in every paragraph, in every chapter, standing out in its most glorious dress.

CHAPTER VII.

For a long time after the hunt recorded in the preceding chapter matters went on in the usual way without anything particularly striking. We at last, however, began to notice signs of an awakening at the F—— appointment. As we stated at an earlier stage in this work, meetings were regularly held in a log shanty roofed with basswood scoops—that is, small logs six or eight inches in diameter were hollowed out in concave form and laid side by side, first with hollow up, then in the second row they broke the joint with reversed side down; this then was the place where services were held, within a few hundred yards of which was a lumber camp containing a foreman and about fifty men.

Our readers will understand that in starting evangelistic services in this place we had an eye

THE OLD F—— APPOINTMENT.
Services were held in the shanty shown in the background.

to reaching the men in camp; and to this end we now seized the opportunity of visiting the place. We were disappointed in not meeting with the occupants the first time, as we had not been made aware of the fact that they took their lunch with them, and remained all day where their work was located, which sometimes was miles away from the camp. The cook, however, invited us to come to dinner at about six o'clock some evening when the men were home. This we gladly did, and received very courteous treatment at the hands of the foreman, who belonged to the Catholic Church. At the conclusion of the meal we asked him if he would object to our having Scripture reading and prayer with the men, and to our delight he said that he would be pleased, and also that he would call them together in the main room, where the beds were arranged, as there would be more space for such a gathering. True to his word, our obliging friend ordered the men to muster in the large room of this low, log building, also roofed with scoops, But some of them, who loved to while away the evening playing cards, telling fictitious stories and singing immoral songs, did not relish attending a religious service; but the foreman was firm as adamant, and we believe it would have cost any man his place if he had absolutely refused to attend the little service that night, so firm was the "boss," as the men termed him.

Of course it must be borne in mind that many of the men were very respectable, and some of them Christians; but as is always the case in a crowd, there were those who loved sinful pleasure better than anything else in the universe. During the progress of the service the best of order prevailed, and at its conclusion we extended a cordial invitation to all present to come to our special meetings.

After conducting them for two or three nights and seeing none of the boys from camp, we decided that they were not going to attend without some further effort being made to induce such a step. Accordingly we found out where they were working in the woods and went over one afternoon, and mingling with the men, set to work pulling the handle of a crosscut saw, which we continued until almost dark, thus accomplishing what reading and praying and everything else had failed to do. Every night after our having helped them the shanty wherein the services were held was literally packed. Some nights almost the entire camp seemed to be there.

Soon a move was made by one or two of the young people accepting salvation. We were troubled a little, however, the latter part of the first week by the misconduct of a few young fellows who sat in the back seat and considerably disturbed the worshippers. For some time we did not like to say anything lest they all take offence and discontinue attending the meetings; but at last it became intolerable, and one night right in the middle of the sermon we stopped, and said, " Now, in the past we have relied upon the honour of those present to conduct themselves properly, and we discovered that they had honour in most cases at least. But if there are any who prove themselves to be utterly destitute of this element, we will at any cost compel them to remain quiet throughout the service."

A large, powerfully-built, rough-looking young man, over six feet high and weighing over two hundred pounds, who was the bully of the entire neighbourhood, leaped to his feet at the close of the above words, and in thundering tones said, " Yes, there must be order, or me and the preacher will lick the whole crowd." It is scarcely necessary to add that the people present, knowing the character of the man, and that he would in a moment back up his words by force if required, gave no further trouble, and from that time forward the best of order prevailed.

The meetings continued for a period of two weeks, and the congregation suffered considerably from the almost unbearable heat. The reader can imagine forty persons or more huddled together in a small room, with ceiling so low that a tall man could not stand perfectly erect, and a large old-fashioned stove with high oven kept almost at white heat. We had indeed a melting time each evening in more ways than one. The result of the services was that about fourteen persons of both sexes professed a change of heart, among them two or three of the men from the lumber camp; one of whom went home suffering from that dread disease consumption, having contracted a cold that finally settled on his lungs while labouring in the woods. We met him on the train when he was going south, and he informed us that although he had doubtless contracted his death in the north, yet he had also found life in the same place.

In recording these evangelistic services and their effect we honestly say that it was not the missionary, nor was it any other mortal creature, that accomplished the above results, but it was purely and simply the work of God.

> Where the word of revelation
> Glows with tidings of salvation,
> Through the cross of Christ made known,
> There thy saving power is shown.

After the winter had passed away, during which the meetings were held at F——, a glorious spring followed with its bright warm sunshine and fertilizing showers, and as this was our last year in Muskoka we determined, although in no wise neglecting our work, to enjoy to the fullest extent the beautiful scenery. The trees were clothed in living green, transformed from the temporary death of winter to the beauteous life of the spring-time, typifying the nature and power of resurrection glory.

From disliking the rocks and the appearance of things generally on our first taking this charge, after having seen more of them, towering up sometimes in perpendicular form to a dizzy height, in other cases framed by the hand of time and by the warring elements into artistic shapes, we learned to love them, for concealed within their flinty bosoms was the recorded history of ages. Also, regarding the forest scenery we may note the fact that it was not only transcendently beautiful in the spring-time, but also surpassingly so in the autumn, clothed in its varied hues of green, amber and gold.

During the month of May we had some inspiring experiences while away on shooting and canoeing expeditions, and truly there is something fascinating in adventure by flood and forest, and the more danger in connection with it the more enjoyable it seemed to be.

On one of our expeditions we fell in with an old hunter who, a little time before, had a very racy experience. Seeing a moose in the woods while on the lookout for a bear, he instantly fired at it, and being a dead shot the bullet took fatal effect, the huge animal running a few steps and falling. Filled with delight at his unexpected success he ran forward and commenced flaying the enormous beast, when a neighbour with whom he had not been on speaking terms for years came along, and taking in the situation at a glance, immediately went to a magistrate and informed on his enemy. The shooter, charged with infraction of the law in killing moose, which was prohibited for some years, was at once summoned, and in his defence swore that while watching for a bear the animal in question attacked him; instead, however, of using his rifle he said that he ran around a large tree, but the moose following, and discerning that his life was in imminent peril, he fired upon it with the above result. His enemy, not seeing the actual killing of the animal, could not, of course, prove otherwise, and the magistrate dismissed the case, advising the hunter to shun the very appearance of evil in the future. The shrewdness of the man caused a good deal of amusement in the community, although all doubtless felt detestation for his lack of integrity. The old saying, "honesty is the best policy," holds good to-day as ever before, and we trust that especially our younger readers will bear in mind that though a policy of deception may seem to be successful for a time, yet that alone will eventually endure and prosper which is based upon solid truth.

And now our sojourn here having ended we took leave of old friends, looking into kind, familiar faces, some of them for the last time, until lake and forest and cemetery shall give up their dead on the morning of the resurrection.

PART II.

VIEW OF PENINSULA LAKE.

PART II.

CHAPTER I.

ONE year later, after receiving ordination, and after linking our destiny to that of one who, amid the difficulties of missionary life, always proved faithful to her trust, we were put down by the Stationing Committee for Dunchurch, in the Parry Sound District. On the way to our new charge we stopped for a day or two at the town of Huntsville, and while there made a tour by steamer of Fairy and Peninsula and Lake of Bays, sailing in one day, according to the testimony of the crew, in the neighbourhood of eighty miles. The trip was delightful, the silvery little seas with their numerous deep bays and green islands and sparkling waters looking exceedingly beautiful.

After the trip just mentioned we passed on to the neat, prosperous little town of Burk's Falls, where we took a substantial-looking steamer with good accommodation, and commanded by Captain K., for A—— Harbour. The trip down the Magnetawan River, through Cecebe and Ahmic lakes, was one long to be remembered. The stream, though narrow, was quite deep, and so winding and with such sharp turns that it required great skill upon the part of the commanding officer, who nearly always took the wheel while in the current, to avoid running into the shore. Such was the efficiency of this functionary, however, that an accident rarely happened.

In many places along the banks the branches of trees touched the water, and all through on either side was the beautiful fringe of green, broken here and there by a clearing, containing farm buildings, which seemed to mar the original beauty of the scene, instead of improving it; thus showing the imperfection of man's work when placed alongside the work of the Divine Architect.

There were one or two wharves—or, more correctly, landings—between B—— and the first lake through which we were to pass, at which the vessel stopped to land passengers or freight, thus giving all who desired an opportunity of going ashore for a time. As they usually took a supply of wood on board at these points, sometimes a whole hour would be at the disposal of those on the vessel.

The first place of importance we arrived at was the village of Magnetawan, which looked respectable and thrifty. At this place were massive locks, built by the Government at heavy cost, and greatly facilitating navigation in these waters. The place also contained four churches, and it was said that three or four ordained ministers resided there; also, as a matter of course, it contained two or three fairly good hotels, four or five stores and a very good Public School building. Immediately after pulling out from the wharf, which was in good repair, a large swing-bridge, also built by the Government, was opened, through which we had the novel experience of passing. In so doing, however, an accident, which might have been very serious, occurred. A Toronto gentleman, wearing a tall and very glossy silk hat, was sitting just in front of the pilot-house, when the mast, which the sailors had forgotten to lower, caught the telegraph wire, giving the bowsprit a sharp jerk, and tilting it backward with considerable force, so as to alight squarely on the crown of our Toronto friend's beautiful castor, which it crushed beyond recognition in infinitely less time than it takes to tell it, also wounding the head slightly, making it bleed profusely. Our fellow passenger was terribly enraged over the accident, through which he lost a first-class hat and which came very near costing him his head. The captain, however, very honourably paid him for the plug, and he showed both wisdom and common-sense in dropping the matter.

The remainder of our journey was uneventful, except that seeing a deer running along the shore of Ahmic Lake caused quite a sensation among those of the passengers who had never before been privileged to see one in its native haunts.

Soon we were in sight of Ahmic Harbour, a neat-looking little place with a nice new frame church, that cost in the neighbourhood of $1,000. Also, the village contained an hotel (which was soon to give place to a very large one, veneered with brick, costing several thousand dollars), a blacksmith's shop, carpenter and paint shop and several private dwellings, one or two of them being very tastily built.

We noticed in all these northern villages an utter absence of those evidences of decay which are far too frequently seen in little towns and hamlets located in the more populous districts near "the front," as the northerners term it.

As we landed we observed a few men loitering lazily around. They looked as though they had quarrelled with work years ago, and had never again made up friends. After some little delay in adjusting luggage, we took our seats in the stage, which was but an ordinary spring waggon with brakes attached—no doubt for the purpose of adding dignity to the vehicle, as there were no hills to amount to much on the road to be traversed. The driver said that it was but four miles to Dunchurch, which was our destination. But we always noticed that in those districts the miles appeared tremendously long, and it seemed to take us an age to reach our final stopping place. At last, however, as we passed over the brow of a bluff, we had a splendid view of the village, which was to be our home for two years.

It contained three churches, one of which had never been completed, two hotels—one licensed to sell intoxicants the other conducted on temperance principles—two very good stores, besides a shoe shop, a saw-mill and sixteen or seventeen dwellings. The parsonage was standing on the side of a hill overlooking Whitestone Lake, and as it was erected on cedar posts, and a very small frame building, it was not very inviting. A ditch had been dug in front of it to drain the road, and the only way of crossing this was to either jump, or, if one was not athletic enough for that, climb down into it and then up the other bank towards the front door.

There was no outhouse on the lot, consequently the horse had to be accommodated in one of the hotel stables, at a cost of nine dollars per month without oats. The expense and unsatisfactoriness of this became intolerable in a month or two, and we determined that a stable should be built. But as the lot was not large enough for this, we first went to the owner of the adjoining land, who gave as much as doubled the size of the parsonage plot without any coaxing on our part. Encouraged by this we at once went around among the people, one hardware man subscribing the nails, a lumber man the scantling and boards, another the shingles, etc. We then announced that work would commence on a certain day, and invited all the able-bodied men, who could do so, to be present to assist.

At the appointed time several were on hand, and for days missionary and people worked together from early morning until dark, the result of which was a fairly good barn, with one stall for horse, driving-room for three rigs, and a loft that would hold three or four tons of hay.

The next thing that required attention was the fact that the village cows loved to hang around the front door and make themselves generally obnoxious. In consequence of this it became evident that a fence around the property was much needed, which was soon constructed, of boards on three sides and pickets in front.

Another thing that needed revolutionizing was the land itself, which was covered with stones and sticks, mortar boxes and debris of all kinds. This foreign matter was soon removed and the place ploughed, and some of it sown with oats, a part planted with potatoes, and three or four beds formed in which were sown beets, carrots and onions. Everything was now getting in shape, and we were beginning to feel comfortable and at home. Trees were also planted along the front, which, if preserved and cared for, would help to beautify the place in coming years.

Our readers will, of course, remember that the gardening was done in the next spring following our settling at D——, the season being too far advanced when we arrived after Conference to do anything in that line.

CHAPTER II.

At the commencement of our pastorate at D—— there were but three appointments, the new church at A—— Harbour, the frame building at D——, and the little log school-house at F——. As we preached in each of these places on the first Sunday of our sojourn there, we were favourably impressed with the intelligent appearance of the people, and also with the fact that they were very respectably dressed. There were scarcely any evidences of poverty, although none of the people were very wealthy. At a former period in the history of the mission, W——, a place twelve miles distant, had been one of the appointments. It lay in an easterly direction from D——, but for nearly a year previous to our being stationed here it had been absolutely discarded; and although the road was very bad in the direction of this place, yet, in response to the earnest solicitation of the people, we re-established regular worship in the frame school-house in that locality, and although the attendance was not large we persistently maintained it, even in bad weather and almost impassable roads, throughout the entire period of our labours in this northern district.

At a later date solicitations came from families residing on the shores of L—— Lake, in another direction, to give them a week-evening service in a new log school-house which they had just completed. As this place was also twelve or thirteen miles away, and the road leading to it in a frightful condition, we at first declined to comply with their request; but they persevered to such an extent that we found it almost impossible to hold out longer, and consented finally to preach to them on Thursday evenings.

One reason for our consenting to perform what seemed to be an almost impossible task was that the spokesman for the people of that locality was an avowed sceptic, and feeling that this might be a providential opening, not only to benefit him but also others who were gradually sliding off in the same direction, we hearkened to the call.

We had but one conversation with our sceptical friend on the inspiration of the Scriptures and the need of salvation, and he wound up by getting very angry, so we were compelled to drop the matter and confine our arguments in favour of both to the desk in the little schoolroom. Mr. F. was a man of more than ordinary intelligence; he had been twenty-one years in the British Royal Engineers, and was in the survey of Palestine and Jerusalem. He was said to be one of the greatest mathematicians in the country. He certainly was a man of noble physical presence, and a fine conversationalist. He was now in receipt of a yearly pension from the Government.

Mr. F. informed us that he went to Palestine a believer and returned home an unbeliever; assigning as a reason for changing his views the fact that at the very doors of the holy sepulchre itself, a regiment of Turkish troops, fully armed, was incessantly on guard to prevent the different sects of Christians from destroying one another. Said he, "If that is religion, I have no use for it." Strange as this may appear to some of our readers, it is the result of ignorance regarding the Scriptures; for within the pages of inspiration it is clearly taught that there will be impostors, wolves in sheep's clothing, false teachers, etc., and if some of them appear before the sepulchre of a slain lamb in the guise of hypocritical fanatics, it is but endorsing the Bible and proving its teachings to be the very epitome of eternal truth.

Like all so-called sceptics with whom it has been our privilege to converse during recent years, our friend showed an inconsistency that at once proved his life to be a deception. He openly avowed his disbelief in Christ or Christianity, and yet his place at our religious services was seldom vacant, and he always compelled his children to go and hear the Gospel.

Although this gentleman was hospitable and intelligent and cultured, a polished, travelled and congenial companion, a faithful and unwavering friend to the servant of God if he believed him sincere, yet away down in his heart he was not what he represented himself to be; and although it has been the lot of the writer to meet many men who have affected to believe even as he, it is our conviction that in all this glorious God-fashioned world there is scarcely a single consistent, honest and genuine infidel.

This appointment at L—— Lake, that Mr. F. was mainly instrumental in persuading us to take up, was continued to the end, in the face sometimes of almost insurmountable difficulties. One night would be so dark that it would be impossible for us to see anything, so that we would be compelled to slacken the reins and let Ned, our faithful horse, pick his way over rocks, and through mud-holes that we were almost surprised did not swallow the entire outfit, and through miles of unbroken forest, our ears regaled with the sound of wild animals leaping out of the roadway into the thicket; another, we would have an upset over a rock, and an almost miraculous escape from a broken neck; again, our cart would break down, or the harness give way; but in spite of all difficulties, we steadfastly persevered, sometimes not getting home until one o'clock in the morning.

We have since learned that the faithful animal that bore us through such trying scenes, though but young in years, died of some disease of the lungs, and we occasionally wonder if the foundation of his malady was not laid upon some of those dreary nights, when in darkness and storm he bore us safely over that awful road; and the marvel is that his master did not go in the same direction.

The first series of evangelistic services held upon this charge were at A—— Harbour, continuing for a period of two weeks. The results, however, were rather disappointing, as only six persons professed to have received a change of heart, nearly all of them young people. An old and devoted Christian lady, knowing that we felt a little dissatisfied because the meetings were not a more pronounced success, informed us that there was a much greater work done than we imagined, and that she herself saw good that had been accomplished where we could not possibly see it, and that eternity alone would reveal it to us.

When the new hotel was erected at the harbour it became a great summer resort, and the following tourist season about sixty Americans were holidaying there for some time. Among them was one man, who, together with his family, represented the cream of United States society; and those who have been so fortunate as to meet with the best people from Cousin Jonathan's dominions, unanimously pronounce them most desirable individuals to associate with.

This gentleman, who was recognized as the leader of the above party, was one of the foremost lawyers in the great Republic, and a member of the House of Representatives, and said to be worth three million dollars. Conspicuous as was his social position, he preached the Gospel as opportunity presented itself. We invited him to preach at D——, which he did with great acceptance, taking for his text, "And Enoch walked with God, and he was not, for God took him." His hands trembled as he stood upon the platform, but he explained that it was not fear that made him shake, "For," said he, "I am not afraid of anything this side of hell."

He then launched out into a beautiful and thoughtful discourse, showing that if we walk with God as Enoch did, we will grow like Him, and His image will be stamped indelibly upon our brow; illustrating it by the work done in the United States mint, where they take the piece of gold and stamp the American eagle upon it and it becomes legal tender for twenty dollars any place in the world. So with the Christian continually communing with God, the stamp of the Eternal is plainly seen upon the life and actions, and he passes for genuine metal of full value any place within this vale of tears.

The latter part of his text, " he was not, for God took him," he shrewdly applied in the following way: If those who profess to believe truly live the Christ life, when it is said of them they are not, God will have taken them to His dwelling-place in the skies. Then he told of a lovely little girl of his own, nine years old, that God had taken, and the first sound he expected to hear at heaven's gate would be the pattering of the little feet over the golden pavement of the eternal city, as she hastened to meet her redeemed parent. As this cultured and evidently sanctified man, in beautiful and eloquent language, portrayed these truths, there was scarcely a dry eye in the crowded church.

Having accepted an invitation to dine at the hotel with this genial and affable American cousin and his family, he gave us some of his experiences during the terrible Civil War, having fought in one of the Union armies. Especially did his eye flash as he told of the awful day at Fredericksburg, when the Federal commander called for a forlorn hope to storm the works. His battalion and two others instantly responded to the call, their leader at once telling them that it was going to be "bloody work"; but they yelled out between their clenched teeth that they would attack the "gates of hell." Then they formed for the charge, and the command rang out, "All ready—forward, at the double," and with a shout that made the very heavens resound they leaped toward the foe, the rebel batteries at once opening on them. But, in spite of grape and canister and shell, and rifle volleys like hail storms, they braved the sleet of death, crossing the open space before the works in three minutes; but oh, horror, what havoc was wrought in their ranks in those awful moments; the ground in front was literally paved with the mangled bodies of their slain comrades. Almost in a point of time many hundreds had fallen, whole companies going down together before the awful tempest of death.

The shattered remnants of the devoted band were compelled to creep under the rebel works and remain there, until in the darkness they were able to make their escape.

CHAPTER III.

SOME one suggested that in the near future it would be enjoyable and healthful to organize an expedition for the purpose of fishing in the Magnetawan river, and as Rev. Mr. D., of Toronto, was in our village in the interests of Sunday-School work, we invited him to join us, which he readily consented to do, being fond of sport, which in no way detracted from his well-rounded character, for he was one of the excellent of the earth. Accordingly next morning we started for the stream, which was said to be an exceptional place for fish, and after walking five miles along the great north road, which led over hills and through forests, we reached the dark waters of this river, noted for being the tomb of many a lumberman and settler. This is accounted for by its great depth and swift current.

Immediately on our arrival we borrowed a punt, a peculiar style of boat of unusual width, having a flat bottom, and commenced trolling just above the rapids, where the stream expanded into a small lake. It required but little skill to fish successfully here, for all that was necessary was to throw the line with hook and feathers attached over the stern of the old scow—which, perhaps, would be a more appropriate name than either punt or boat—and then row for the rapids.

There was some danger in connection with this, however, as the nearer we approached the rocks and foaming waters the better we found the fish to bite, and the reader will readily understand that the temptation is exceedingly great to get as near as possible; and a plunge over the seething torrent would mean good-bye, fishermen, until the judgment.

We, however, continued pulling back and forth, every few minutes hauling out a fine fish, until at last we had quite a number, some of them weighing several pounds. After each one was drawn in there was a shout of victory that made the welkin ring, and at last, after one of the finest day's sport that mortal ever enjoyed, we returned the old boat to its owner with thanks, and two or three good fish thrown in; and loaded with spoil walked once again the five tedious miles, reaching the parsonage about dark, and as you may imagine doing ample justice to the viands spread before us.

Mr. D. and the writer were accompanied on the above occasion by Mr. E., who was a very remarkable character. He was the father of a family of nine daughters, although his partner was now many years dead. The earlier part of his life was characterized by great wickedness. Indeed, he was a confirmed drunkard until after he had passed the meridian of life. He was now, at the period of our fishing expedition, living upon borrowed time, that is, he had passed the three score and ten milestone.

He of whom we write was once so lost to all sense of shame and sunken so low as to walk through the main street of the village in a perfectly nude condition. This, of course, would not have been allowed but for the absence of the constable. Some considerable time after this had happened he attended a series of special services held on the charge, and was soundly converted, from that time forward engaging in every good work with astonishing alacrity and zeal, being finally licensed by the Quarterly Board as an exhorter. While taking appointments he would sometimes drop his studied theme and relate his own experience with the tears streaming down his face, a convincing proof of his sincerity. In this manner he helped others into the kingdom of God, and the closing years of his life became productive of good.

This strange person has since passed away to his eternal rest, and after an intimate acquaintance with him, extending over a period of two years, we believe that he has gone to the paradise of God to sing the new song, the first and last of the celestial hymnal, "Worthy is the Lamb that was slain."

A long period was now spent in pastoral visitation and ordinary routine mission work, giving once again admirable opportunities of studying human nature in its varied phases, and also of forming more correct conceptions of the habits of the people and mode of life in the newer districts of Canada. Some of the settlers were very bright and intelligent, and others were ignorant and stupid. We do not know that they differ in this way from the general population of the country, whether urban or suburban; in all cases there are the brainy and those who are conspicuous alone for lack of this element. In reality society is divided into two great classes—not the rich and poor, not those of aristocratic birth and those of humble parentage; but the patrician or true nobleman, with intellect and piety, and the plebeian, possessing neither the one nor the other. This classification is, we believe, on the principle of true democracy.

We are pleased to say, after mingling with our countrymen in many parts of this great Dominion, that the intelligent largely predominate.

As the season advanced we imagined that there were signs of a revival breaking out at the D —— appointment. Accordingly, we announced for special services to begin the following week, and for half a month they continued, with very little success. The very heavens seemed brass and the earth iron; and, altogether, during the continuance of those meetings only four individuals professed to have received a blessing equivalent to conversion.

Wherever we have had failure we have always candidly acknowledged it, or at least made an honest attempt to do so; and with regard to this series of meetings of which we now write, so far as numerical results are concerned, they were a failure. But there is one redeeming feature, if nothing else, in connection with them, that causes joy, which is the fact that one of the four converts, a young school teacher, is now preaching the Gospel as an ordained minister in connection with the work of one of the great evangelical denominations. We trust that he will eventually bring to a glorious consummation the work so feebly begun.

After the strain of the above services we

started out very early one morning in quest of deer, thinking that the best rest was a change of occupation for a day or two. We plunged through tangled underbrush, and over fallen trees, until at last we had traversed miles of forest, and could find scarcely a living thing, not even a chipmunk. Having taken lunch with us, we spent the whole day in the woods, and sometimes coming on fresh tracks and following them up our hopes would rise, only to be shattered again as we would reach the shore of some small lake to find that his deer-ship had taken to swimming, and was now far out of reach of Winchester bullets.

As the shades of evening began to gather, we were reminded that we must at once retrace our steps, or have the questionable delight of spending the night in the woods. In our efforts to get out of the seemingly interminable forest and reach the main road, we began to realize that we had covered a good many miles more than was intended. This is easily understood when the reader considers the eagerness with which the hunter follows in the track of an animal that has passed perhaps less than five minutes, and is every moment expecting to get a fair shot.

We have been told by old hunters that they have frequently followed wild beasts so far, and so eagerly, that they have been compelled to camp where night overtook them, sometimes in dangerous places frequented by wolves.

Having found the road, however, we reached home about nine o'clock at night, weary and footsore and hungry, but otherwise none the worse for the day's sport. Hunters learn to call everything fun, no matter how much hardship there is in connection with it.

An old settler informed the author that on one of his deer-stalking expeditions, he was following a runway so earnestly that he did not notice darkness coming on, and with his eyes continually on the tracks, which were fresh and numerous, he never for a moment dreamt of obstructions being in the way, when suddenly looking up he found that he had almost run against a large black bear with two cubs. The

TROPHIES OF THE CHASE.

animal at once showed fight, and promptly received a bullet for her temerity. But unfortunately the rifle was a single muzzle loader, and Mrs. Bruin was only made furious by the wound received; she at once dashed at him, and he said, "If ever mortal man got a hustle on" (using his own language) "he was that man." He sprang into a tree that was close at hand, but quick as he was the bear caught one of the legs of his pants as he swung himself up, tearing a portion of it clean off. It was truly a close call, and our hunter friend informed us that he was very thankful to escape into the tree with the loss of a part of his trousers, and of course his rifle, which he dropped when making the frantic effort to escape the embrace of his new acquaintance.

Speaking of the experience afterward, he said that he believed every man had a chance sometime in his life to accomplish something, if he had been provident enough to keep the means to that end within his reach. He, however, sinned away his day of grace when he dropped his gun, and as he sat on a limb near the top of the tree in a chilling wind for two mortal hours, with one of his lower limbs covered, and at least half of the other in a perfectly unprotected condition, watching a wounded and angry bear, he would have given worlds, if he possessed them, for a "shooting iron," as he termed it, to practise for a minute or two on mother Bruin and her progeny; but as he beheld the weapon in which he implicitly trusted lying harmlessly on the ground, and the little cubs playing with it, he became so furious that he kicked the branch on which he sat with his heels, and in trapper's phraseology, swore in a manner that would fill his satanic majesty with delight. Finally the trio at the foot of his uncomfortable home grew weary of watching and waiting for one last fond embrace, and disappeared in the dark woods. Our friend promptly descended from his lofty perch, gathered up his scattered belongings and started for home, arriving at which he almost threw his wife into hysterics by his dilapidated and woe-begone appearance. Feeling a little ashamed of the whole performance he did not care to say very much about it for a long time afterward, but at last began to regard it as a good joke, and talked quite freely of the adventure so fraught with danger.

CHAPTER IV.

DURING the early part of our first winter in D—— we started out on a missionary tour, many miles to the north, in the direction of the great lumber camps. Our work consisted of baptismal services, Gospel talks, and the visitation of isolated families living beyond the bounds of our own mission. In doing this we were not trenching on the rights of any sect or denomination, for those people never entered a Christian church, except, perhaps, on the occasion of a funeral, and then they would have to drive a long distance over a road that simply beggared description.

It was pleasant, in a certain sense, to visit them, for although they did not care very much for religion, yet they always seemed to appreciate a visit from a minister of any persuasion. The writer was called into one home, where were the father, mother and two children —one of the latter having been baptized the preceding week by Rev. Father ——, a Catholic priest—and was requested to administer the ordinance of baptism to the other, which we cheerfully did, it being the most remarkable experience of our lifetime, for as we took the child in our arms, to perform the rite, it became very angry, and scratched our face and pulled our hair in vigorous style.

During this Conference year of which we write, we were called upon to apply the water to fifty-nine children—fifty-two being the children of Protestant and seven of Catholic parents. Some of them were beautiful, clean, bright little things, and others badly needed the application of water.

In some cases, at least, we are strongly in favour of baptism by immersion, as the only bath that some people ever take is when they are dipped; and if this process, which is but symbolical of soul-cleansing (as likewise is the sprinkling of a few drops upon the subject), will not wash away sin, it will at least wash away moistened dust particles, and is thus productive of good. We are also so liberal-minded as to believe in the final perseverance of the saints. The only difficulty which is to be deplored is that many of them do not finally persevere, and consequently get into mischief. And concerning apostolic succession we are very enthusiastic, knowing beyond all doubt that the labours of those upon whose heads were laid a Redeemer's hands, and who from a Saviour's lips received at once their commission and credentials, "Go ye forth into all the world preaching the Gospel to every creature," preceded by many centuries the work of active Christians of to-day, no matter of what persuasion; thus coming after or succeeding those early teachers and preachers, the truth stands forth, clear as the mid-day sun, that we are all as God's workmen in this bright succession.

With regard to the lumber camps in this most northerly point that we had yet reached, we may say that they were run, some of them, on an unusually extensive scale, one in particular, containing an hundred men, the output of logs from it during the winter being enormous. It was operated by an American company, with a capital, it was said, of six million dollars.

Our much-loved friends of the great Southern Union come to us, in some cases at least, for their timber supplies, as their resources in that particular direction become exhausted, and as ours are almost inexhaustible. If our Republican cousins should ever find that all their supplies had utterly failed, and if in consequence they should desire it, we would be pleased to take them under our protecting wing, and annex the United States to Canada, changing slightly the paragraph in our Public School geography containing the boundaries to the following: " Bounded on the north by the Aurora Borealis, on the south by the Gulf of Mexico."

When the men returned in the spring-time from the woods, and turned their horses out in the corral at A—— Harbour, it was a grand sight to see so many together. It looked as though an entire cavalry regiment had dismounted and let their horses loose for a roll and feed of grass. The headquarters of the Company was at this place, where they had a very comfortable office, also buildings for the storage of supplies and stabling on a very extensive scale, with a large staff of employees attached.

On one of our missionary journeys we were told of a very peculiar character who lived in a little clearing of an acre or two, that he had made with his own axe, and which he planted from year to year with potatoes. He held very strange religious views, and bitterly hated preachers of all denominations, having insulted every man of that stripe who had ever dared to call upon him.

Although we were warned not to go near this dangerous individual, as he was termed, yet, like our lady friends, the warning not to look or see only increased our curiosity, and that to such an extent as to determine our going. We became extremely desirous of seeing this person, and hearing what he had to say. In advancing toward the door of his little log-house we scarcely knew what to expect, whether a furious and hungry dog would be let loose on us, or the contents of a shot-gun emptied into our epidermis, or lastly and least probable, a fairly cordial welcome be extended. Knocking at the door, which was half open, we waited in breathless suspense for a response. At last some one said, in a very gruff voice, "Come in," and doing so we found ourself in the presence of a tall, lank, lantern-jawed individual, fully six feet three in his stockings. Although there was a good deal of curiosity behind this visit, yet in the main we felt we were about our Master's business, and consequently put on an unabashed front.

When seated he at once began setting forth his peculiar views, and of course nobody, whether Protestant or Catholic, could agree with him. After each statement he would say, "Now, Mr. Preacher, isn't that so?" and although we hated to differ with him, because of his unsavoury reputation, yet even Satan could not fall in with his ideas, and we had either to lie, putting it in plain terminology, or contradict him. On our doing so at last, however, he got very angry, and said he did not believe in salaried ministers, as they were only disturbing the people in the midst of a peaceful trust, and were a lot of miserable imposters. We paid no attention whatever to this last charge, believing that the manifestation of a proper spirit at this time was the most effective argument; and so it proved, for he soon settled down to a normal condition, and talked quite calmly and respectfully.

The noon hour having arrived, the visit, contrary to our custom, being made in this particular case in the forenoon, we were invited to remain for dinner, and as things did not look very clean we bitterly regretted having talked so long. But there was now no escape, as this man with whom we conversed knew very well that there was not another family living within two or three miles, and we could frame no reasonable excuse.

The head of the house remarked that they had very little in the place, but such as they had the preacher was most heartily welcome to share with them. There being no other alternative we sat down to table with the lord and lady of the establishment and their ten children, the eldest being about ten years old. The table contained nothing but potatoes, butter and milk, and being urged to eat and drink heartily we tried to show our appreciation of everything by doing the best we could. Drinking the milk was no punishment on ordinary occasions; but on this, what with foreign matter floating upon the top, and a very strong odour arising from the centre, and a vast accumulation of dust and other things, which we could not analyse without a glass, in the bottom, it was quite a different matter. We also tried very hard to choke down a couple of tubers, but it was sorry work, as they did not seem to taste right without some accompaniment, and we must confess that for some time after the meal we felt a little squeamish. We next had prayer, with the permission of the host, followed by another argument on the Scripture, and then took our leave, inviting the household to attend worship at the nearest shanty, in which we preached every two weeks to a few of the settlers.

We noticed one thing, however, in our visit, that whereas this peculiar character could not provide proper food or clothing for his numerous progeny, yet he could provide himself with both whiskey and tobacco; his breath evidenced the former, and he proved his devotion to the latter by smoking right in our presence. We have no war with the man who can discharge his duty toward his family, his fellowmen, and the Church of God in every particular, and who, after he has done all this, occasionally indulges in a smoke. But we have a war to the very knife with those who will neglect their families because of needless self-indulgence, and who will either absent themselves altogether from all manner of religious services, or be paupers in the sanctuary. They are, we believe, the most deadly foes to Christian civilization and to society, and in the sight of God the greatest sinners this side perdition.

With regard to our friend of whom we have written, we may add that another visit was soon made to his home, and he commenced attending. divine service soon after, became interested at last, began praying in public, was soon a devoted member of the Church, and a respected and patriotic citizen. Having discontinued his evil practices he became more prosperous, and soon his family were much better fed and clothed than ever before. The last time we dined with them we had a really pleasant time, and they not only had the milk, potatoes and butter as of old, but also eggs and bread. It had taken many weary months to get Mr. F. into line. We often thought that some other brother would have had him inside the spiritual kingdom of God in half the time; but hard as the work had been, and long the time consumed in doing it, we felt amply rewarded by seeing so great a change in his life.

On taking leave of the above person for the last time before quitting the place, we ventured to ask him if he disliked preachers as much as formerly. He threw his head back and laughed heartily, as he said that he had really met but one that he could not get along with, and the thought had often come into his mind recently that he had been as much to blame, perhaps a good deal more, than the missionary, and he thought that on the whole they were not a bad lot of fellows after all.

In depicting character in this, as in all other cases, we have earnestly endeavoured to be true and faithful in every particular; and as the same conditions are to be found everywhere, we have in nowise cast a reflection on the people of our northern districts. There, as in the south, and in the south as there, much of the squalor and misery that exists is the result of the liquor habit, which induces idleness and shiftlessness.

It must be remembered, however, that there is a vast difference between that condition which is the result of intemperance, and that which is known as honest poverty. And if any of our readers have not sufficient sympathy for the honest poor, we would suggest that they spend a few years in the mission field, and there they will learn that poverty is no evidence of vulgar birth or lack of culture; for many of those who are at the present time in that condition are, according to our classification of society in a former paragraph, true patricians— that is, they are to be found in the ranks of God's noblemen. Whereas, some who are differently situated are of vulgar birth and questionable lineage, having been raised to affluence by a "corner" on pork or malt, or something else, and are plebeians of the commonest type.

In thus stating what every thinking man, whether rich or poor, will endorse; we do not cast a single shadow upon any class, for we love all alike, believing in the fatherhood of God and in the brotherhood of man, and in the truths of Christian inspiration, which teach that we must not despise the day of small things, nor hold in contempt the "simple annals of the poor."

CHAPTER V.

AFTER all the varied experiences recorded in the last chapter, we come to one of a widely different character, namely, the revival at F——. For some time past there were evidences that the seed was taking root, and finally, one Sunday morning, in the short after-service someone stood up for prayer, seeing which we at once announced for special meetings to commence next evening. The active members promptly rallied around the missionary and faithfully discharged their duty, and in consequence of this diligent toil for the salvation of men, at the close of the first week there were several professing regeneration, evidently with great sincerity.

On the following Sabbath the place was packed almost to suffocation, and was dreadful, like Bethel to Jacob, because of the presence of the Lord. We have always noticed that wherever we undertook to force a revival, failure followed. But, on the other hand, if it broke out spontaneously, it was always a crowning work, proving that all man can do is to sow the seed; everything else is of the Lord of Hosts. Seldom, we believe, does a series of evangelistic services come to a termination without Satan making an attempt to frustrate the efforts of God's people. Twice with insidious cunning was this designed by the arch enemy: First in stirring up a spirit of sectarian bigotry, as there were four or five different denominations represented; second, to employ a half-drunken hotel-keeper to run into the preacher's rig on a dark night with a heavy team and spring waggon, and pile cart, horse and missionary all in a heap in the fence corner, which actually happened.

He was defeated in the former case in the following manner: Immediately on our finding out that there was going to be friction we plainly told the people to get converted first, which was before everything else, and then they could join any Church they desired; stating still further that we did not ask any of them to connect themselves with our society—that was a secondary matter, the all-important thing was to become children of the heavenly King.

The above declaration, which embodied the true sentiment of our soul, completely disarmed the cavillers, and the work went on gloriously unto the end.

In the latter case his failure came through the preacher not being killed in the smash-up caused by his agent, who, at the time, was half filled with distilled damnation. Having recognized the wretch who had caused the damage we at once waited upon him next morning, and informed him that he was liable to the extent of the injury done the cart, which consisted of five or six spokes being knocked out of one wheel, the dashboard broken, etc. He cheerfully consented to make it good, which he promptly did, also apologizing for having endangered our neck, and expressing deep regret at the unfortunate circumstance throughout. Our almost miraculous escape from death or broken bones was, in part at least, due to the faithfulness of the noble animal attached to the rig, who, although knocked over by the fearful shock, instead of springing to his feet and kicking or running away, and finishing the work of demolition, cautiously rose and stood stock still until everything was patched up temporarily, and we were able to go on to our appointment.

On reaching the little school-house we found it filled with eager worshippers, and as our clothes were torn and we were smeared with blood and dust, we were under the necessity of telling the astonished people what had happened, and some of the young men present made open threats of violence on the individual who had proved himself such a miscreant. A night or two later as we lay upon our couch, after the first sleep, perhaps about twelve or one o'clock, some drunken men, driving at a furious pace, passed the parsonage yelling like fiends, "Show us the man that run into the preacher." And we must confess that for two or three days we were afraid that some injury would be done to the person or property of the publican; and after his repairing the damage and apologizing we did not desire this—indeed, we may add that we would not countenance reprisals under any circumstance.

The affair recorded above, instead of being an injury, gave a fresh impetus to our meetings, and each succeeding night afterward many were found seeking the Saviour, some with terrible earnestness, and in nearly every case proving the truth of the Scriptures as they record the immortal words, "They that seek me shall find me." Altogether the meetings continued for two weeks and a half, when heavy rains put a stop to them.

About seventeen persons professed to have accepted the Saviour; and although the number does not appear large, yet it must be remembered that it comprised at least one-third of the entire audience and, with one or two exceptions, all the unrestrained sinners in the community. Of course, in candour, we must add that some of these lapsed, but then it should be borne in mind that this is always the case; a certain percentage will, as time goes by, relax their hold of a new found Saviour, and in consequence drift away. Perhaps in those cases the work has not been properly done. Whereas, on the other hand, we feel safe in saying that the majority will maintain inviolate unto the end the faith once delivered unto the saints, finishing their course with joy and receiving their everlasting reward amid the plaudits of assembled worlds.

CHAPTER VI.

THERE were two households in the F—— neighbourhood that professed Christianity, and who were diametrically opposite in every particular, giving a true picture of human nature within the pale of the Church. The C. family, the first of these, were brought up in the fear of the Lord, and taught by devoted Christian parents to hate sin in every form and to love righteousness. After such a training need it surprise any that they grew up to be a benediction to society in their own community, with characters stable and solid as the rocks that surrounded their humble home. The family consisted of mother, father, and ten children, and was among the most prosperous in the district. This temporal success was due in large measure to the thrift and industry which was the outcome of genuine religion, and altogether these remarkable people were an example along the lines of sterling integrity to the entire local population. To give an illustration of the fortitude and devotion of the mother, even in death, we quote the following from a young preacher, who was intimately acquainted with her the last year or two of her life: "She arose in the morning, apparently in her usual health, which was never perfect, and said to her family, 'I do not know how it is, but never before has this text of Scripture been so strongly impressed on my mind, 'Be ye also ready, for in the day or hour ye think not of, the Son of man cometh.' Sometime during the afternoon she said to her daughter, 'I do not feel well, help me to my room,' and as they were entering it, she whispered, 'I am ready,' and a few minutes later was not, for God had taken her."

Thus our readers have in the picture of this family a crowning example of true godliness in life, and of triumph and victory in the death of one of its members.

"Oh, may we conquer so,
  When all our warfare's past,
  And dying find our latest foe
  Under our feet at last."

In the second instance, the F. family made the same profession as the other, but they were shiftless and indolent, and in consequence of this never had the necessaries of life in sufficient measure, to say nothing about any of the luxuries. The house in which they lived was dilapidated-looking outside, and inside so filthy that it was almost incredible that any human creature could exist in it. If this were all, perhaps the picture would not appear so unsightly, but to aggravate the situation the head of the establishment frequently went to the adjoining village to do business, and sometimes returned therefrom in an intoxicated condition, making things lively at home on his arrival. The children were disobedient, ignorant, half-naked, and wretchedly dirty.

The two portrayals are now before our readers. On the one hand are an honest confession of faith in Almighty God, and lives that are in harmony with that profession, and an example of victorious trust and dying exultation to prove its beauteous reality.

On the other is hypocritical pretension, with all the train of woes that follow in its heaven-forsaken wake, such as drunkenness, beggary, filth, squalor, misery, and, too often, consummate ignorance, without one single golden bow of promise to light the gloom of earthly night or loose the grip of satanic power.

It is now for our readers to choose between the real article, with profession and life to correspond, and the spurious, reeking with the fumes of the pit of despair that can never strike a harmonious note.

Soon after the F—— revival we were asked to conduct the sacramental services at the village of M——, in a church which was truly founded upon a rock, and judging from the spirit of piety manifested by the people on that commemorative occasion, the gates of hell had never been able to prevail against it. The following day we were asked to baptize a lady of about forty years, who became troubled in her mind concerning the matter, fearing that if death came her chances for future happiness would be doubtful. When it required so little to make her happy we unhesitatingly complied with her request, and administered the rite by pouring. The lady referred to was in delicate health, which accounted in large measure for her anxiety relative to her unbaptized condition.

A few weeks later we were asked to conduct quarterly services at McKellar, a village lying in an opposite direction; and after a pleasant and, we trust, profitable morning service, a number of us took a trip down the lake in a steam launch to an afternoon appointment, where we found a large number of people awaiting us, and where we had a rousing meeting, followed by a eucharistic service, and that by a good supper at a neighbour's, immediately after which we were compelled to hasten back to McKellar for the evening service.

The return trip up the lake was one long to be remembered. When about half-way we passed through narrows which barely allowed room for the steamer, with a scenery on either side the depicting of which would utterly baffle the pen of the most gifted writer. Here an enormous rock, there a clump of trees with their many-tinted foliage, sometimes almost representing the matchless colours of the rainbow, while yonder could be seen a farm with carpet of living green. On arriving at our destination we found the church crowded, and after a service which personally we enjoyed very much—we cannot answer for the congregation, whether they experienced enjoyment or endured suffering—we proceeded to the hospitable home of Mr. A., where, surrounded by every comfort and entertained by himself and his amiable and intelligent partner, we soon forgot the toils of the day.

While recording these pleasing experiences, we must not forget a visit to the Parry Sound camp-meeting, where many hundreds of people were assembled and where the Gospel was delivered with soul-convincing power. A novel scene was the delivery of a sermon by Rev. Mr. S., through an interpreter, to the Indians, one or two hundred of whom were present. More striking still was a morning prayer-meeting held solely by the hardy sons of the forest, and though we could not understand their words, as they spoke in their mother tongue, yet we could judge of their sincerity by the intense earnestness manifested in gesture and look. They seemed to be natural orators. The camp ground was well fitted up with all the appliances needed for the convenience of those who attended, also furnishing everything necessary for the successful carrying on of evangelistic work. Mr. B., at whose home we stopped during the short period of our visit, took a very great interest in the good work of saving men. Large hearted, cultured and refined, with a sympathetic nature, he loved his fellow-creatures, and none rejoiced more than he when souls were swept into the spiritual kingdom of God. This gentleman had one of the choicest families that it has ever been the privilege of the writer to become acquainted with; mother and children were ornaments to society and a benediction to the Church of Christ.

During our visit to the Sound we formed a most favourable opinion of our red-skinned brethren, and we believe that if they had enjoyed the same privileges as those of fairer complexion they would have been as well qualified to hold positions of trust and honour as the best of them. We have examples in Church and State, in more than one land, where Indians who have acquired a liberal education have taken a prominent and conspicuous place, and in our own Dominion we are proud to have in our midst the original proprietors of the soil, now fellow-countrymen of copper hue, who would, we believe, if the need existed, show their patriotism and valour on any field.

CHAPTER VII.

DURING the autumn of our last year in the P—— S—— District we wrote to Rev. Dr. Shaw, Assistant Missionary Secretary, to come and give us a couple of good strong addresses in the interests of the Society. He at once responded, and we met him at A—— Harbour; but having our light Armstrong cart along, the venerable preacher was slow to enter it, saying that "he did not like the look of those machines." Seeing, however, that the roads at that particular season were almost impassable for a four-wheeled vehicle, he at last consented to take his seat, and before we reached D—— he was quite in love with the conveyance. The springs being single leaf, the motion of the cart was very pleasant, and almost before we had time to realize it we had arrived at the parsonage. As the meeting was not until eight o'clock

REV. JOHN SHAW, D.D.

there was ample time for conversation and story telling, and those who knew the genial Doctor best were well acquainted with the fact that he could tell a capital story. As some apology was made for us failing to find in any of the stores a particular luxury designed for the guest of the evening, that gentleman at once said that it reminded him of a Governor who was visited by a certain functionary, and who, according to custom, should have had a salute fired from the battery. This he did not do, however, and the visiting officer asked him his reason for the omission of this mark of respect and courtesy. Said he, in reply, " I did not order the firing of the guns for twenty-one reasons, which I shall proceed to enumerate. First, I had no powder. Second,— "That will do," said the officer, "I accept the explanation."

We do not know whether in this case our welcome and honoured visitor thought that we were going to give twenty-one reasons for the absence of the article, but on our saying we had visited all the stores and found that none of them had it in stock, he said that was sufficient. Immediately on the arrival of the hour for our missionary service we wended our way to the church, where a goodly number had assembled to hear the distinguished speaker, who had not proceeded far with his address until all could see that he was at his very best.

It had been the privilege of the writer to hear the lecturer of the evening on more than one occasion; but the oration delivered at this gathering surpassed in power anything that we had ever heard from the Doctor's lips. He captured the hearts of the people by his opening sentences, when he told them that in his younger days he had frequently sat at the foot of a tree waiting for daylight, that he might commence chopping. The reader can imagine the effect of this upon men who were themselves woodmen, and the flashes of wit and humour, together with a peroration of surpassing eloquence, preserved the attention of the entire company to the end.

The following morning we invited our venerable friend to accompany us on a deer hunt, and as he was quite experienced along this line, and just as much at home in the chase as the writer, he joyfully accepted the challenge. We borrowed a good double-barrelled shot gun from our Recording Steward for the Doctor, also extending a very hearty invitation to Mr. T., the above official, who was a man of excellent parts, to accompany us on the expedition. Owing to some previous engagement, however, he was unable to accept, and expressed his surprise and delight at the enthusiasm of the hoary-headed, sociable and fluent divine, who started out on the hunt with all the eagerness of a school boy.

Our armament consisted of a sixteen-shot Winchester repeating rifle, and the double shot gun which we had borrowed for the occasion, the latter loaded with grape. We had scarcely gone a mile when a fox crossed the road about two hundred yards ahead, and having a small field of two or three acres to pass through before reaching the woods, he was fully exposed to our fire, which was delivered without ceremony. Crossing the field had brought him within about one hundred and fifty yards of where we stood, and although he ran very rapidly, yet by allowing for his speed, and the time a bullet would take to reach him, we had done enough shooting to know about how far to fire ahead of Mr. Reynard, as he ran broadside to us; and when the fusilade was over, although a few bounds more would have landed him in the woods, safe from the leaden hail of his enemies, he lay upon the grass motionless in death, stricken through the heart. Although neither of us delighted in taking the lives of innocent and harmless birds or animals, a reprehensible practice, yet we felt on this occasion that we had rendered a service to the settlers by the destruction of an animal that had robbed them of many a good meal, by taking their fowls, which some of them could ill afford.

After disposing of our prize, we proceeded some miles farther, a part of the way following the tracks of a large deer of the male gender, which can always be determined by the condition of the front part of the hoof, which in the female is always sharp, and in the other rounded. We followed the trail for a long distance through the dark unbroken forest, and at last, to our humiliation, reached the shore of a good-sized lake, only to find that his deer-ship had taken to the liquid element and was now far away. Determined, however, to have some sport, we fixed up a target against a tree, and practised shooting for some time. The Doctor proved himself an excellent marksman, and no doubt Reynard's death was due to his deadly skill.

As the afternoon was wearing away we found that it was necessary to at once find our way out of the woods, which we would have experienced some difficulty in doing without a compass, but for the fact that our companion, with his ripened experience, knew the different points by the moss on the trees. On the way home our most entertaining fellow-traveller told us a story of an Irishman, who had just come to this country, and was taken by some friends for a hunt. He was placed on a leading runway, with a gun in his hands, and told to shoot the first thing that came along. When the hounds had ceased running, the other members of the party gathered around our friend, the son of Erin, and said, "Patrick, why did you not shoot the deer?"

"Oi niver saw anything of that name!" said the representative of the Emerald Isle.

"But," said they, "it passed right here!" showing him the fresh tracks.

"Well, gorrah, I niver saw any deer, so I didn't," said Pat; "but I saw the divil, with a rockinchair on his head and a white handkerchief on his tail. He ran roight past me, so he did; and froightened me so bad, that oi couldn't shoot, so I couldn't."

Early next day Doctor Shaw, who had during his short stay at our home greatly endeared himself to us, bade us an affectionate farewell and took his departure. Little did we think, as we looked upon that benign countenance that was never darkened by a frown, that we would not see it again upon earth, yet this was to prove a sorrowful fact. Owing to a lamentable accident, caused by a trolley-car in the city of Toronto, this able and consecrated servant of God has since passed from earth away. One of the most intelligent laymen in the Province of Ontario, in speaking of him who has gone from us, said that "John Shaw was in every respect a Christian gentleman. The translation of this gifted man of God has left our western hemisphere poorer, but has increased the wealth of the paradise beyond."

> "Servant of God, well done !
> Thy glorious warfare's past ;
> The battle's fought, the race is won,
> And thou art crowned at last."

Although truly loyal to the Church of our choice, yet in a spirit of modesty we have refrained from recording its name in these pages. Those most devoted to the cause which they have espoused can best afford to be liberal-minded, and we trust that our readers, whether Catholic or Protestant, as they peruse the lines within this work, containing an abbreviated narrative of a few of the events of the lives of three worthies who belonged to one of the greatest Christian denominations upon the American continent, having representatives in every quarter of the globe, and their translation from the Church militant upon the earth to the Church triumphant in the heavens, will resolve that they will mould and fashion their lives and characters more after the pattern of the God man, who left His footprints on the sands of time, crimsoned with His life's blood. For truly He who framed the universe with infinite skill, and fashioned worlds from naught, was the example loved and followed by those whose names we have reverently recorded.

> "Thus star by star declines,
> Till all are passed away,
> As morning high and higher shines
> To pure and perfect day."

CHAPTER VIII.

WE have in the preceding chapters endeavoured faithfully to record the memories of the past relative to the Canadian mission-field, and we now judge it pardonable if we break through the boundaries of our northern districts and deal with a few facts concerning our beloved country generally, without repeating any of the comments made upon its resources and scenery in Part I. of this volume. The area of the entire Dominion is 3,205,343 square miles, and although our fellow-citizens sometimes complain of the fact that our population is not increasing as rapidly as we might hope, yet impartial statistics prove that it has multiplied more rapidly by many per cent. than the population of the great Republic south of us. Even our American friends acknowledge this, and though the disparity in numbers between these two neighbouring States is so great, it must be borne in mind that it was much greater one hundred years ago; and we venture to predict, that if there were less partyism and more true patriotism, our vast Dominion would prosper even more. It was said of ancient Rome, that she rose to the zenith of her power when none were for a party and all were for the State.

At a banquet given by the Toronto Board of Trade some time ago, a spirit of robust Canadianism of sentiment and optimism of view was manifested respecting the future. It was a mark of praise in an ancient patriot, that he never despaired of his country in the most adverse circumstances. So not only the Queen's representative, but also the leaders of both political parties, united in the laudation of Canada, in confident anticipation of its brilliant future. The following figures were given by His Excellency, the Governor-General, for four years of financial depression, namely, from 1888 to 1892. During the first year the total imports were $110,894,000; in 1892 they had risen to $127,400,000. The exports in the same time had risen from $90,000,000 to $113,000,000. Also the sea going and lake tonnage had, during the quadrennium, risen from $15,000,000 to $19,000,000; whilst the coast trade tonnage had increased from $18,000,000 to $25,000,000. The total increase in employed registered shipping was from $34,000,000 to $43,000,000. During the above four years of depression in trade and agriculture the deposits in the Government and special savings banks had risen from $182,000,000 to $229,000,000.

Since the year 1892, although a great financial strain has been felt in almost every clime, and we Canadians have ourselves suffered to some extent; yet on the whole, considering every phase of the question, our native land has made marked progress. Our shipping is increasing, and in this year in which we write the waters of the ocean are being ploughed by the new steamship *Canada*, of nine thousand tons register, equipped in the most modern fashion, and being the first vessel of a fast Atlantic line, destined, we believe, when all the vessels are completed, to be the admiration of the world.

Also our defences are being strengthened, and the work of rearming the militia with the new Lee-Enfield magazine rifle, one of the most effective weapons ever devised, is progressing.

In spite of trade depression, the amount of money deposited in our banks is continually on the increase, and altogether, under the able adminstration of the Earl of Aberdeen, who is now Her Majesty's representative in this country, there is ceaseless progression along the lines, which is cheering to every loyal heart. Our increasing commerce is incontrovertible evidence of this.

As the result of the general elections a change of Government has taken place during the last summer, with the Hon. Wilfrid Laurier at its head, a man in many respects a typical Canadian; and now almost at the dawn of another year, which brings us nearly to the threshold of the twentieth century, after watching his actions as Premier for some time, we feel hopeful that the man whose hand is on the helm of State will prove as great a statesman as the late

Sir John A. Macdonald, who was revered and honoured by Canadians everywhere.

With regard to the population of our country, we can say, as did an American Senator of his: "That already we hear the tramp of teeming millions as they swarm up our smiling valleys and over our rugged hills, hewing out homes for themselves and families in this wondrous land of freedom and liberty, destined, we believe, under the providence of God, to some day reach the one hundred million mark."

The growing feeling to-day is not for annexation to the United States, much as we may respect that enterprising people, but is, we believe, for a closer federation with the much-loved Mother Land. As a non-party man, we quote the words of Sir Oliver Mowat, when he said: "As a Liberal, and knowing something of the Liberal mind, I affirm that neither party is an American party, that both are essentially Canadian, and that however they differ otherwise, both are, as parties, opposed to giving up our half of the continent to the Southern Republic, and openly antagonistic to thereby blotting out forever the name of our beloved Canada."

We believe that there is on this continent more than ample room, and more than sufficient resources for two great nations. There are, no doubt, individuals here and there in every province who desire political union with our neighbours, but the great mass of the Canadian people do not favour it, and never have.

Then, enjoying, as we do, the privilege of practically ruling ourselves, and cultivating a national Canadian sentiment, irrespective of creeds or parties, let us be loyal first to our God, second to our country; and then as Atlantic and Pacific billows dash with sullen roar upon our eastern and western shores, may it be but the musical cadence that shall stimulate us in our onward march to take our place amid the Parliament of Nations, and in the federation of the world.

The End.

www.ingramcontent.com/pod-product-compliance
Lightning Source LLC
Chambersburg PA
CBHW022118160426
43197CB00009B/1078